Enzyme Studies

H. M. Kalisz

Microbial Proteinases

M. Maestracci, K. Bui, A. Thiery, A. Arnaud, P. Galzy

The Amidase from a Brevibacterium Strain: Study and Applications

AKADEMIE-VERLAG BERLIN

Enzyme Studies

Enzyme Studies

Managing Editor: A. Fiechter

With 20 Figures and 24 Tables

Akademie-Verlag Berlin 1988

Die Originalausgabe erscheint im Springer-Verlag Berlin Heidelberg New York
als Volume 36 der Schriftenreihe
Advances in Biochemical Engineering/Biotechnology

Vertrieb ausschließlich für die DDR und die sozialistischen Länder Akademie-Verlag Berlin

ISBN 3-540-18997-1 Springer-Verlag Berlin Heidelberg New York Tokyo
ISBN 0-387-18997-1 Springer-Verlag New York Heidelberg Berlin Tokyo

ISBN 3-05-500568-6

Erschienen im Akademie-Verlag Berlin, DDR-1086 Berlin, Leipziger Straße 3—4
Lizenznummer: 202 · 100/532/88
Printed in the German Democratic Republic
LSV 1315
Bestellnummer: 763 885 3 (3070/36)
10800

Prof. Dr. Bland Symington Montenecourt

Bland Symington Montenecourt

Bland Symington Montenecourt, Professor of Biology, Lehigh University, died of cancer Saturday, December 19, 1987.

Bland received her B.A. in biology, magna cum laude, from Rosemont College in 1964. While attending Rosemont, she achieved national ranking in tennis and played against the great women tennis stars of the 1960's. From Rosemont College she went to Rutgers University, receiving a Ph.D. in Microbiology in 1968 working under the guidance of Joel O. Lampen on the regulation of enzyme synthesis in yeast. After a two year post-doctoral appointment at Rutgers Medical School, Bland took time off from her active career to raise a family — two sons and a daughter.

In 1976, she joined the Department of Biochemistry and Microbiology, Cook College, Rutgers University. There, as an Assistant Research Professor, in collaboration with Douglas E. Eveleigh, she established a world-renowned program in the isolation of hyper-cellulolytic strains of *Trichoderma reesei* for use in the fermentative conversion of biomass into industrial chemicals.

Bland came to Lehigh as Associate Professor in the Department of Biology in the Fall of 1981. She was promoted to Professor in 1985. During her tenure at Lehigh she continued her interest in biomass conversion, investigating the genetics of *Trichoderma reesei* and *Thermomonospora sp.* cellulase production and, more recently, expanding her program to include the study of thermophilic clostridia, the microbial fermentation of cheese whey, and the microbial desulfurization of coal.

Bland was a prolific researcher. In the past five years she authored or co-authored 52 publications. She was an international leader in the development of microbial strains for the fermentative conversion of biomass into industrial chemicals. She was a consultant for several national and international companies. She served on the editorial boards of four biotechnology journals including *Applied Microbiology and Biotechnology*, *Trends in Biotechnology*, *Journal of Biotechnology*, and *Applied and Environmental Microbiology* and not to forget *Advances in Biochemical Engineering/Biotechnology*.

In recognition of her many accomplishments she was awarded the Annual Research Award of the Division of Microbial Chemistry and Technology of the American Chemical Society, the Novo Award for Excellence in Research, and the Eleanor and Joseph F. Libsch Research Award from Lehigh University.

Through her dedicated and prominent involvement in professional activities, she served as a role model for students. She encouraged their active collaboration with visiting scientists and solicited their involvement in proposal writing and grant management.

Bland integrated her professional career with numerous outside interests, serving a an active member of the church and on the Bethlehem Town Environmental Commission. She had a love for animals and was a farmer/scientist who had experience with and understood the use of steam exploded wood as a ruminant feed for her cows, goats and sheep. She was a nurturing parent and took particular pride in her three children — Dorsey, Ned and Marc. Their support and that of her husband, Gene, sustained her through her illness and allowed her to work under increasingly difficult circumstances.

We will remember Bland for her great activity and forthright manner. Most of all we will miss her as a good friend and caring, valued colleague. May she rest in peace.

Janice A. Phillips
Associate Professor
and Arthur Humphrey, Doug Eveleigh and Kathy Gottlund

Table of Contents

Microbial Proteinases

Henryk M. Kalisz
Gesellschaft für Biotechnologische Forschung (GBF) mbH, D-3300 Braunschweig, Mascheroder Weg 1, FRG.

Advances in Biochemical Engineering/
Biotechnology, Vol. 36
Managing Editor: A. Fiechter

Up to 10 years ago proteinases were regarded as degradative enzymes which could only catalyse the total hydrolysis of proteins. However, recent advances in assay techniques, such as the use of more selective substrates, have demonstrated that proteolytic enzymes carry out highly specific and selective modifications of proteins by limited hydrolysis. Furthermore, proteinases play an important role as reagents in laboratory and clinical analyses and in industrial processes. Thus, considerable emphasis is placed on research on the proteinases. The discovery of new, highly specific proteinases and improved enzyme technology, such as immobilisation and novel purification methods, should make the microbial enzymes even more attractive in biotechnology. Development and modification of existing industrial systems is likely to be the main factor in increasing the industrial application of the proteinases. In genetic biotechnology proteinases have mainly a detrimental effect in protein purification and the elimination of these enzymes may be an important factor in potential progress in this field. This review describes some of the specific functions and properties of the microbial proteinases and discusses some of the most important commercial applications of these enzymes.

1 Introduction

Since their introduction in the 1960's as detergent additives, the commercial usage of proteinases has progressed rapidly. As the most important industrial enzymes, proteinases account for nearly 60% of total enzymes sales [1] (Table 1), with two-thirds of the proteinases produced commercially being of microbial origin. The most important microbial proteinases employed commercially are the alkaline proteinase of *Bacillus licheniformis* used in detergent manufacture; *Mucor* proteinase used in cheese manufacture, and *Aspergillus oryzae* proteinases used for dough modification and soy sauce production. These enzymes account for the bulk of proteinase production world-wide. However, other proteinases, both microbial and non-microbial, are also employed in numerous industrial processes. This is especially true in the food industry where fungal proteinases are of particular importance.

Table 1. Total worldwide enzyme sales

Enzyme		Total Sales	(%)
Proteinases	Alkaline (Detergents)	25	
	Rennins	10	*59*
	Trypsin	3	
	Others	21	
Carbohydrases	Amylases	18	
	Isomerase	6	
	Pectinases	3	*28*
	Cellulases + Lactases	1	
Lipases			*3*
Others	Analytical		
	Pharmaceutical		*10*
	Developmental		

2 Detection of Proteinases

Most proteinases are normally detected qualitatively by using proteins as substrates. Proteins such as haemoglobin, casein, azocoll, and Remazoll Brilliant Blue have been used extensively. Proteolytic activity is usually detected by either;

1) Following the decrease of the initial substrate, or
2) Measuring the increase of defined peptide products. In many cases proteolytic activity is followed by measuring the release of peptides from dye-, fluorescence-, radioactively-, or enzyme-labelled proteins. This simplifies the measurement of peptides and increases the sensitivity of the assay.

Activities of proteinases, such as the blood clotting enzymes, kallikreins and many other mammalian proteinases, which catalyse the hydrolysis of specific peptide bonds are usually measured using the natural substrates of the enzymes.

Detection of highly specific proteolytic enzymes has advanced considerably with the introduction of synthetic peptide substrates for the assay of enzyme activity.

Suitable peptide portions, located on the amino-terminal side of the bond to be cleaved, are used as the specific recognition sequences. Hydrolysis of the selected peptide releases a chromogenic (e.g. 4-nitroaniline) or fluorigenic (e.g. 7-amino-4-methyl coumarin) compound, which can subsequently be measured spectrophotometrically or fluorimetrically, respectively.

In addition to detection of proteinases by specific substrates they can be identified and characterized by affinity labelling methods. Compounds such as DFP, TLCK, TPCK, peptide chloromethyl ketones and peptidyl diazomethyl ketones have been used to distinguish between individual proteinases present in biological samples.

Details of the above-mentioned methodological approaches are described elsewhere [2-6].

3 Proteinase Functions

3.1 General Characteristics

Proteinases are a highly complex group of enzymes which vary enormously in their physico-chemical and catalytic properties. The proteolytic enzymes, which are produced intra- and extracellularly, play an important role in the metabolic and regulatory processes of animal and plant cells, as well as in those of prokaryotic and eukaryotic microorganisms.

Extracellular proteinases are involved mainly in the hydrolysis of large polypeptide substrates, such as proteins, into smaller molecular entities which can subsequently be absorbed by the cell [7,8].

Intracellular enzymes play a key role in the regulation of metabolic processes [9,10]. They also play a vital role in protein turnover [10,11], maintaining a balance between protein synthesis and degradation [10,11]. Proteinases are also involved in the control of many other physiological functions, such as digestion [12,13], maturation of hormones [14-16], viral assembly [17-19], immune response [3,20-22], inflammation [23-25], fertilisation [23,26], blood coagulation [3,23,27], fibrinolysis [3,23,28], control of blood pressure [23,29-31], sporulation [15,32,33], germination [13,34,35] and pathogenesis [36,37]. Proteinases have also been implicated in the regulation of gene expression [35,38], DNA repair [39,40], and DNA synthesis [41].

3.2 Protein Turnover

Protein turnover is a continual process in all living cells [10,42,43]. The process is similar in all organisms, with different rates of turnover for individual proteins and subcellular fractions [35,44]. Protein turnover eliminates abnormal proteins [45], such as mutant human haemoglobins [9], and is essential for adaptation of cells to new environmental conditions, especially in response to starvation for nutrients [10,15,35]. Degradation of a protein appears to be initiated by exposure of a unique cleavage site at the C-terminal end of the protein [46].

Breakdown of polypeptides of no value for the cell provides a pool of amino acids as precursors for the resynthesis of proteins. Under conditions of starvation, proteins serving newly required functions in the cell can be synthesized with little net change in protein content [47,48].

Evidence for the involvement of intracellular proteinases in protein turnover is available [15,30,49]. ATP-dependent proteinases of *Escherichia coli* and resting yeast cells are responsible for the hydrolysis of the abnormal proteins [50,51]. In *E. coli*, the energy required for the degradation of these proteins is due to proteinase La, the *lon* gene product [52,53], which hydrolyses proteins and ATP in a coupled process [54]. The rates of protein degradation in vivo depend on the cellular content of the ATP-dependent proteinase, with enhanced proteolysis being deleterious to the *E. coli* cell [55]. Proteinase La, a 450,000 Daltons tetramer composed of identical subunits [53], catalyses an initial rate-limiting step in abnormal protein degradation in vivo [54]. Energy-dependent proteolytic enzymes other than proteinase La also exist in *E. coli* [56,57]. The energy-dependent degradation of proteins is a multi-step process involving activation of the proteinase by the binding of ATP; cleavage of the peptide bond; and hydrolysis of the ATP to ADP which stops protein cleavage [58]. Protein degradation appears to require two molecules of ATP per peptide bond cleaved [59]. Hydrolysis of ATP, however, is required only for the degradation of proteins; cleavage of peptides of less than 10,000 Daltons requires only binding of the nucleotide to the enzyme [58]. In eukaryotes, the highly selective non-lysosomal turnover of intracellular proteins is also effected by ATP-requiring mechanisms [51], involving a ubiquitin-dependent pathway [51,60]. The degradation process requires ATP for the binding of ubiquitin to the substrate protein [61]. Degradation of the ubiquitin-protein conjugate also requires ATP and may involve ATP-dependent proteinases [62].

3.3 Sporulation

Conditions of starvation induce the formation of bacterial spores [35,63], yeast ascospores [64], and the fruiting of slime moulds [13] and fungi [65]. These processes involve intensive protein turnover [35]. Proteolytic processes play a key role in the transition from vegetative growth to spore formation in yeasts and bacteria [15,33,66].

During the aggregation of unicellular slime moulds into multicellular units and subsequent differentiation, sweeping changes in enzyme patterns occur, with a large amount of protein breakdown for the synthesis of new proteins and carbohydrates [67]. Large increases in proteinase activity are also observed during fruiting body formation in the cultivated mushroom *Agaricus bisporus* [65] and the related "bracket fungus" *Schizophyllum commune* [68]; the proteinases in the latter being involved in enzyme inactivation reactions.

In bacteria, at least two intracellular serine proteinases are activated by a post-translational event at the end of vegetative growth and during the early stages of sporulation [69]. A membrane bound proteinase, which shows unique properties and is produced earlier than the cytoplasmic enzymes during sporulation [70,71], controls the proteinase system in the sporulating cells of *Bacillus subtilis* by regulating the level of a proteinaceous inhibitor which specifically inactivates the cellular proteinases [72]. Production of extracellular proteinases also coincides with sporulation [73], the increase in activity resulting from de novo synthesis. Experiments with proteinase inhibitors have shown a need for both extracellular [32] and intracellular [74] proteinases in sporulation. However, *B. subtilis* mutants deficient in the two major extracellular proteinases sporulate normally, indicating that these sporulation-associated enzymes are not essential for spore development [75,76]. Proteinase inhibitors have also been used to show the need for proteinases in sporulation of yeast cells. Protein breakdown during yeast sporulation is energy dependent. In *Aspergillus* species a correlation between proteinase secretion and sporulation has been reported. The region of proteinase formation is at a constant distance from the hyphal tip, and has a constant spatial relationship with the membrane site where the first conidiation septum is formed [77].

3.4 Germination

During bacterial spore germination, proteinases make amino acids available for the synthesis of new enzymes required for the process and provide a nitrogen source for nucleotide biosynthesis [35]. About 20% of the protein in dormant spores of *Bacillus* and *Clostridium* species is degraded in the first minutes of germination [78]. Proteolysis is necessary for protein synthesis by the germinating and outgrowing spore, since the dormant spore lacks most free amino acids and is incapable of de novo synthesis of a number of amino acids due to an absence of the necessary enzymes [79]. Degradation is initiated by endoproteinase activity of at least two serine enzymes [34]. Both proteinases have unique specificity for storage proteins, which are a group of small, acid-soluble spore proteins synthesised only during sporulation under transcriptional control [80,81] and do not affect other spore proteins, and are present only in the developing forespore. They are absent from vegetative and young sporulating cells, and from the surrounding mother cell. Activity of both enzymes is rapidly lost on germination [34].

Proteinases are also involved in macroconidial germination and hyphal fusion. Macroconidial germination in *Microsporum gypseum*, and an alkaline proteinase found in the germination supernatant, can be inhibited by a specific serine proteinase inhibitor. Addition of the proteinase increases spore germination [82]. Following hyphal fusion between *Podospora anserina* strains which are heterogenic for their incompatibility alleles, an incompatibility reaction occurs which is correlated with the release of high proteolytic activity [83]. During germination of *Dictyostelium discoideum* spores [84] and *Polysphondylium pallidum* microcysts [85], excreted acid proteinases are thought to be involved in the breakage of cell wall polypeptide linkages.

3.5 Enzyme Modification

3.5.1 Activation

Many inactive precursors of zymogens are converted to active forms by proteinases which cleave one or more peptide bonds. The proteinases involved in such reactions are usually highly specific. Zymogen activation represents an important part in physiological regulation, being a rate-controlling step in many processes such as the generation of protein hormones [15,16], activation of enzymes [12], assembly of fibrils [86] and viruses [17,18], blood coagulation [23,27] and fertilisation of ova by sperm [87].

Only a few examples of enzyme activation in lower eukaryotes have been observed. In yeast, proteinase contained in vesicles activates a zymogen of chitin synthase by limited proteolysis in vitro [88]. The active chitin synthase is involved in primary septum formation in budding yeast cells [89]. Activation of chitin synthase has also been observed in *Phycomyces*, *Candida albicans*, *Mucor rouxii*, *Aspergillus nidulans* and related species [89,90]. In vitro activation of the *Mucor rouxii* cAMP phosphodiesterase has also been demonstrated [91]. The vacuolar enzymes of *Saccharomyces cerevisiae* are also activated [15,92]. Proteolytic processing of prohormones occurs at sites containing paired basic residues [93]. The processing of α-mating factor in *Saccharomyces cerevisiae* appears to involve two novel proteinases; a serine proteinase in the α-cell soluble fraction [14]; and a calcium-dependent cysteine proteinase in the α-cell membrane fraction [94].

3.5.2 Inactivation

Inactivation is defined as an irreversible loss of in vivo catalytic activity in the physiologically significant reaction of an enzyme [47]. It is distinct from protein turnover, where proteins are degraded to their constituent amino acids. In vivo enzyme inactivation has frequently been observed in microorganisms in response to physiological and developmental changes [47]. Several enzymes are irreversibly inactivated after a metabolic shift. Enzymes involved in gluconeogenesis, such as phosphoenolpyruvate carboxykinase [95], fructose-1,6-bisphosphatase [96] and cytoplasmic malate dehydrogenase [97] become inactivated when glucose is added to *Saccharomyces cerevisiae* cells growing on acetate and the mechanism of inactivation involves proteolysis.

Proteinase B from yeast has been shown to inactivate several enzymes in vitro, such as tryptophan synthetase, chitin synthase, fructose-1,6-biphosphatase and glutamate-oxaloacetate transaminase, while yeast proteinase A inactivates threonine dehydratase and tryptophan synthetase [98]. However, studies with mutants lacking these two proteinases have shown that neither enzyme is solely responsible for inactivation [15,33]. In the inactivation of fructose bisphosphatase, proteinases are involved in the second of a two stage inactivation process which starts with a reversible phosphorylation [99]. The enzyme appears to be proteolysed by extravacuolar proteinases [100]. The proteolytic degradation of a protein has been shown to depend upon the conformational stability of the molecule [101]. Highly specific degradation of proteins may be achieved in a two-step process involving the covalent modification of the proteins as a marking mechanism for proteolysis [43].

Many enzymes of vegetative cells are selectively inactivated during or prior to sporulation. Hydrolysis of anabolic enzymes not needed during the stationary phase may make available peptides and amino acids for other purposes, such as spore synthesis or formation of new enzymes necessary for adaptation to the changing growth environment [102,103]. Most examples of inactivation have been found in bacilli and yeasts. They include NADP-dependent glutamate dehydrogenase, isocitrate lyase, as well as the gluconeogenic enzymes, malate dehydrogenase and fructose-bisphosphatase, whose activities decrease from the onset of yeast sporulation [103], and aspartate transcarbamylase [104], glutamine phosphoribosylpyrophosphate amidotransferase [105], and threonine dehydratase [106], which are inactivated in starving

B. subtilis cells prior to sporulation [48]. Inactivation of aspartate transcarbamylase either consists of, or is immediately followed by, selective proteolysis, in an energy-requiring process [107]. Inactivation of glutamine phosphoribosylpyrophosphate amidotransferase requires oxygen and is followed by proteolysis [108]. Several enzymes, including the glutamine synthetase of *Klebsiella aerogenes* and *Escherichia coli* become proteolysed following covalent modification by mixed function oxidation [43, 109]. The glucose dehydrogenase of *Bacillus megaterium* is cleaved into fragments by proteinase K, resulting in a loss of activity [110]. In *Neurospora crassa* tryptophan synthetase is inactivated by a proteinase during transition from exponential to stationary phase [111]. In *Schizophyllum commune* iso-enyzmes of phosphoglucomutase are selectively inactivated during fruiting by a proteinase which is only detected during inactivation of the enzyme [68].

3.5.3 Modification

A number of enzymes are modified so that their physiological reactions become altered but not lost. Leucyl-tRNA synthetase from *Escherichia coli* is modified from an enzyme catalysing synthesis of leucyl tRNA to one which catalyses leucine-dependent pyrophosphate-ATP exchange, by a proteinase which splits a polypeptide of about 3000 molecular weight from the native synthetase [112]. The fructose bisphosphate aldolase of *Bacillus cereus* [113] and the RNA-polymerase of *B. thuringiensis* and *B. subtilis* [114] become modified by proteinases, probably during transition from vegetative growth to sporulation.

Enzymes modified by proteinases in yeasts include 3-phosphoglyceric acid mutase, hexokinase, aldehyde dehydrogenase, phosphofructokinase and cytochrome b2 [98]. These modifications could, however, be in vitro artifacts, and may play no physiological role in vivo.

3.6 Protein Maturation and Secretion

Many proteins are synthesised as precursors and are subsequently processed by limited proteolysis to the mature (authentic) products. Four protein maturation events involving proteolysis have been recognised [33]. These are:

a) Removal of N-terminal formylmethionine or methionine from nascent polypeptide chains;
b) Removal of peptide extensions or "signal peptides" from proteins passing through cell membranes;
c) Cleavage of the translation product of polycistronic mRNA coding for several distinct polypeptide chains;
d) Conversion of inactive pro-proteins into biologically active products, i.e. enzyme activation.

Proteins translocated across or integrated into cellular membranes are synthesised as precursors containing a short amino terminal extension of 15–30 amino acid residues, called "signal peptide" [115]. At least three structurally and functionally distinct regions have been defined; a positively charged basic amino-terminal region containing 2–8 amino acids (n-region), followed by a 7–15 residue long uncharged, mainly hydrophobic region (h-region), and a 5–6 residue long polar

C-terminal region (c-region) [116]. The amino acid before the cleavage site has a short side chain [117]. Prokaryotic proteins have a considerably higher incidence of acidic than basic residues around the cleavage site [118]. The signal peptides are essential for the translocation of proteins across cell membranes [117]. They are recognised by the signal recognition particle (SRP), a cytoplasmic ribonucleoprotein, which binds to ribosomes and may cause arrest of elongation of the initiated polypeptide chain at a discrete site. The SRP-ribosome complex binds to the SRP receptor, or "docking protein", an integral membrane protein of the endoplasmic reticulum [119]. The SRP is released on binding, the ribosome resumes chain elongation and the growing polypeptide chain passes through the membrane [120]. The signal peptides are cleaved off during or shortly after membrane passage to yield the mature product [117,121,122].

3.7 Extracellular Proteinases

Extracellular proteinase activity has been observed in numerous prokaryotic and eukaryotic microorganisms [7,13,123,124]. Extracellular proteinases are involved mainly in the degradation of exogenous proteins to peptides and amino acids before cellular uptake. They usually have wide substrate-specificities and can degrade most non-structural proteins, such as albumin, casein, insulin or haemoglobin. Extracellular proteinase activity has also been observed during sporulation [73] and spore germination [85]. Synthesis of the *B. subtilis* serine proteinase appears to be regulated either by a catabolite repressor or an inducer produced only at the end of logarithmic growth [125].

Many pathogenic microorganisms secrete proteinases, some of which are involved in the infection process [13,126,127]. The virulence of a few pathogenic fungi and bacteria, such as *Vibrio cholerae* [128], is correlated with extracellular proteinase activity of the organisms. The proteinases of viruses have also been implicated in infectious diseases [129]. Several species release specific proteinases which can hydrolyse structural and connective tissue proteins resistant to attack by most proteinases. These include the collagenases of *Clostridium histolyticum*, *Pseudomonas aeruginosa* [13] and *Lagenidium giganteum* [130]; the elastases of *B. subtilis*, *Ps. aeruginosa* and *Flavobacterium*; and the keratinases of *Streptomyces fradiae* [13]. Some species, such as the dermatophytes *Microsporum* and *Trichophyton* produce all three types of proteinases. A number of bacteria, including *Neisseria gonorrhoeae* and *N. meningitidis*, *Haemophilus influenzae*, *Streptococcus pneumoniae*, *S. sanguis* and *S. mitio*, responsible for various human infections release IgA1 proteinases, which specifically cleave the immunoglobulin IgA1 which normally provides for antibody defence of mucosal surfaces [36]. All of the bacterial IgA1 proteinases tested are metalloproteinases [131], except for the *Bacteroides melaninogenicus* enzyme, which is a cysteine proteinase dependent on divalent cations [132]. Several pathogenic *Candida* species secrete acid proteinases which can degrade IgA and which exhibit keratinolytic or collagenolytic activity [133].

Many pathogenic species, including the apple pathogenic fungus *Monilinia fructigena*, utilise host proteins for nutrition [134]. Some species release several proteinases, each with different metabolic functions. One such is the parasitic flagellate

Trypanosoma cruzi, which secretes three proteinases which are believed to be involved in nutrition, parasitism and escape from the host's immune response [135].

4 Localisation and Control of Proteinases

4.1 Localisation of Intracellular Proteinases

Many organelle-specific proteinases have been observed [136]. Proteinase activities have been found in nuclear chromatin [137] and ribonucleoprotein particles, in mitochondria [138, 139], chloroplasts, and ribosomes. Signal peptidases, catalysing the removal of signal peptides from secretory proteins have been observed in rough microsomes. Membrane-bound acid proteinases have been found in rough and smooth microsomes of *Aspergillus oryzae* [140]. A number of enzymes have been found in *Escherichia coli* cell membranes [141]. In eukaryotic microorganisms proteinases are contained in membrane-surrounded compartments similar to the mammalian lysosomes [98, 142], such as the vacuoles of *Saccharomyces cerevisiae*, *Neurospora crassa*, *Candida albicans* and *Microsporum gypseum*. Proteinase activity has also been found in large vacuoles of starved *Euglena gracilis* cells and in digestive vacuoles of various flagellates [143]. In prokaryotes, proteinases are present in the cytosol and are connected with the membrane fraction [141].

4.2 Control of Proteinase Activity

4.2.1 Intracellular Proteinases

4.2.1.1 General Mechanisms

A number of mechanisms operate to control proteolysis. These include modulation of substrate proteins by covalent interconversion, change in hydrophobicity and interaction with various molecules, which change their susceptibility to proteolysis [51, 136]. Proteinase activity is also controlled by nutritional conditions and catabolite repression. The proteinase activity increases in both microbial and mammalian tissues under conditions of nutrient starvation [9]. Nutrients such as glucose repress proteolytic activity in yeasts, bacilli, *Escherichia coli* and other microorganisms [98, 136].

Control of proteinase activity can also be effected by limited proteolysis, as in the activation of zymogens and processing of proteins [15, 92, 144], and by specific localisation of the enzymes. Low molecular weight effectors, such as ATP [9, 51], divalent cations, especially calcium [145], and charged tRNA [146] are able to modify proteinase activity. In the case of charged tRNA, guanosine tetraphosphate and other guanosine nucleotides are thought to serve as allosteric effectors of the proteolytic system.

4.2.1.2 Control by Inhibitors

An extensive number of proteinase inhibitors of microbial origin has been observed [147]. Many of the inhibitors are active against a large number of proteinases of similar properties. Others, such as those from yeasts and *Neurospora crassa*, are very specific and inhibit only one type of endogenous proteinase or peptidase. Yeast contains three inhibitors which specifically inactivate proteinase A and B and carboxypeptidase Y [148]. Proteinase A and B inhibitors are heat and acid resistant. Two forms of inhibitor A and three of inhibitor B have been found, exhibiting differences in their isoelectric points. With the exception of the third proteinase B isoinhibitor, which has a molecular weight of 11,5 kDa, the different forms have around 8 kDa. The carboxypeptidase Y inhibitor is heat and acid labile and has a molecular weight of 23,8 kDa. The inhibitors bind to the enzymes in a 1:1 ratio [13,90]. Proteinase A and B inhibitors are inactivated by proteinase B and A, respectively, and carboxypeptidase Y inhibitor becomes inactivated by both proteinases A and B. Carboxypeptidase Y has no effect on any of the inhibitors. The inhibitor concentration is always in excess of the proteinase and both activities increase markedly on transition from logarithmic to stationary phase [90].

N. crassa contains at least four heat-stable inhibitors with molecular weights between 5 and 24 kDa. Two of the inhibitors are specific for aminopeptidases, the other two preferentially inhibit serine proteinases of *N. crassa* and yeast. The inhibitors are located in the cytosol; the enzymes are present in the vacuoles. A similar intracellular distribution has been found in *Aspergillus nidulans*. The inhibitors are thought to protect the cell from proteolytic damage which might result from leaky or broken vacuoles. They may also be involved in the control of both general protein turnover and specific limited proteolysis [90].

4.2.2 Extracellular Proteinases

Four possible mechanisms of control of extracellular enzyme production (synthesis and secretion) have been suggested [8]:

a) Constitutive synthesis with regulated secretion, where the enzyme is synthesised constitutively but its secretion is regulated by induction and/or repression. This mechanism has not been demonstrated for extracellular proteinases.
b) Constitutive synthesis and secretion, where the enzyme is constantly secreted into the medium regardless of the substrate. This mechanism is not known to exist in fungi.
c) Induced synthesis with constitutive secretion. De novo synthesis of the enzyme is induced by the substrate, or a metabolically related molecule. Proteinases may be induced by proteins and large polypeptides. In cells unable to store newly-synthesized extracellular enzymes, control of secretion by induction cannot be distinguished from control of synthesis.
d) Derepressed synthesis with constitutive secretion. Proteinase production is regulated by derepression through an absence or limitation of a carbon, nitrogen and/or sulphur source, in the absence of an inducer.

Evidence exists for proteinase induction and derepression, as well as catabolite repression. Proteinases of the bacteria *Micrococcus caseolyticus*, *Serratia* species and *Vibrio parahemolyticus* are induced by peptides or proteins. These enzymes appear in the media during exponential growth. Synthesis of the extracellular proteinase of *Micrococcus freudenreichii* is increased by certain amino acids, and the *Serratia* enzyme is induced by leucine [149].

Many bacteria produce extracellular proteinases maximally during the stationary phase, probably to utilise macromolecules in the environment in response to nutrient limitation [150]. Proteinase synthesis is also regulated by catabolite repression, by easily metabolisable sources of carbon and amino acids [149]. Production of extracellular enzymes during the stationary phase is thought to be controlled at the transcription level [151]. In some bacteria, such as *Pseudomonas* and *Arthrobacter*, production is constitutive.

In fungi, as in bacteria, extracellular proteinases are produced by many species from each of the major taxa [13, 123]. Although much information is available about the properties of the extracellular proteinases, little is known about the mechanisms controlling their production. Most of the work on the mechanism of biosynthesis of extracellular proteinases in fungi has been from studies on *Aspergillus* species [152, 153] and *N. crassa* [154, 155].

In *Aspergillus* species, proteinase production is controlled by derepression in the absence of a carbon, nitrogen or sulphur source. The organisms secrete at least four proteinases, three of which are stable and which are functionally different from the intracellular proteinases. Derepression involves de novo synthesis. The *Neurospora* enzyme is controlled by induction and repression. As in *Aspergillus*, induction involves de novo synthesis, and the secreted proteinases exhibit different properties from the intracellular enzymes. Two of the proteinases, one alkaline and one neutral, are regulated coordinately and are induced when protein is present with one of the nutrilites absent [154]. Approximately 0.4% of the total protein synthesis is devoted to the formation of the alkaline proteinase, which is a glycoprotein [156]. Three acid proteinases are released under varying conditions of derepression [157]. In the presence of protein and all three nutrilites proteinase production becomes repressed. Peptides and amino acids do not effect induction, but certain amino acids inhibit proteinase synthesis when added to media deficient in one of the nutrilites. In *Neurospora* the neutral and alkaline proteinases produced under the three limiting conditions are identical, which suggests the enzyme is regulated by a single structural gene in a complex fashion, and is activated by each of the three distinct metabolic signals. Expression of the structural gene is regulated by a *nit*-2 gene, which along with certain minor structural genes control the production of a number of nitrogen metabolism enzymes. The product of the *nit*-2 gene has been identified as a DNA-binding protein [158]. In both *Ascomycetes* species maximum rates of proteinase biosynthesis occur long before appreciable cell growth.

In the basidiomycetes production of extracellular proteinases appears to be regulated mainly by induction. The enzymes are not repressed by the nutrilites and are apparently bound to the mycelial wall during the early stages of growth [159]. The basidiomycete *Coprinus cinereus* secretes at least five proteinases of molecular weight 31 kDa; two cysteine, two metallo-, and one serine proteinase [160].

5 Properties of Microbial Proteinases

The classification system for microbial proteinases, recommended by the Nomenclature Committee of the International Union of Biochemistry [161], divides the proteolytic enzymes into two major groups — peptidases and proteinases — on the basis of their nature of attack. The exopeptidases remove terminal amino acids or dipeptides and are of secondary importance since they cannot rapidly complete digestion. They are sub-divided further according to whether they act at the C or N terminal (carboxy- or aminopeptidases), or on a dipeptide (dipeptidases). The carboxypeptidases and the proteinases, which cleave internal peptide bonds, are sub-divided further according to their side-chain specificity, as proposed by Morihara [162]. Proteinases are classified by their catalytic mechanism into four groups. This classification is determined indirectly through reactivity toward inhibitors which react with particular residues in the active site region. The four groups are 1) serine proteinases (EC 3.4.21); 2) cysteine proteinases (EC 3.4.22); 3) aspartic proteinases (EC 3.4.23); 4) metalloproteinases (EC 3.4.24).

5.1 Serine Proteinases

The structure and catalytic mechanisms of the serine proteinases are described in detail [12,162,163].

The serine proteinases are the most widely distributed group of proteolytic enzymes of both microbial and animal origin [13,87,123]. The enzymes have a reactive serine residue in the active site and are generally inhibited by either DFP or PMSF. Many are also inhibited by some thiol reagents, such as pCMB, probably due to the presence of a cysteine residue near the active site which probably does not participate in the catalytic mechanism of the enzyme [13]. Serine proteinases are generally active at neutral and alkaline pH, with an optimum between pH 7–11. They have broad substrate specificities, including considerable esterolytic activity toward many ester substrates, and are generally of low molecular weight (18,5–35 kDa). However, the largest serine proteinase reported is the *Blakeslea trispora* enzyme, with a molecular weight of 126 kDa [164]. Most have isoelectric points between pH 4.4 and 6.2.

Serine proteinases can be divided into four sub-groups, according to their side chain specificity against oxidised insulin B-chain [162].

5.1.1 Trypsin-Like Proteinases

Trypsin-like proteinases are produced by a number of *Streptomyces* species, such as *S. erythreus*, *S. fradiae* and *S. griseus*. These enzymes are specific for basic amino acids, are most active at pH 8 and are sensitive to trypsin inhibitors, DFP, soybean trypsin inhibitor and TLCK. Their molecular weights are about 20 kDa and, with the exception of one of the *S. erythreus* proteinases which has a pI of 4, have isoelectric points at pH 9 [162].

5.1.2 Alkaline Proteinases

Alkaline proteinases are specific for aromatic or hydrophobic residues, such as tyrosine, phenylalanine or leucine. They are produced by various species of bacteria, moulds and yeasts, and are sensitive to DFP and a potato inhibitor, but are unaffected by specific trypsin inhibitors TLCK and TPCK. They are most active around pH 10. Their molecular weights are in the 15–30 kDa range, and their isoelectric point is normally around pH 9.

The best known of the alkaline serine proteinases are the Subtilisins produced by *Bacillus licheniformis* and related bacilli (Sect. 6.2.2.1). Other alkaline serine proteinases are produced by *Arthrobacter*, *Flavobacterium arborescens* [165] and *Streptomyces* species. Alkaline proteinases are also produced by fungi, such as *Aspergillus* species, *Neurospora crassa* [166], the thermophile *Malbranchea pulchella*, and *Saccharomyces cerevisiae*, whose alkaline serine proteinase is thought to be involved in the processing of prohormones in vivo [14].

5.1.3 Myxobacter α-Lytic Proteinase

This enzyme, produced by a species of *Sorangium*, exhibits strong bacteriolytic activity toward a number of other soil bacteria. Like elastase, the enzyme hydrolyses polypeptide chains with a preferred specificity for peptide linkages involving the carboxyl groups of neutral, aliphatic amino acids. The proteinase exhibits maximum activity at pH 9 and is inhibited by DFP [162].

5.1.4 Staphylococcal Proteinase

A *Staphylococcus aureus* strain produces a DFP-sensitive proteinase with a molecular weight of 12 kDa and maximum activity in the pH range 4.0–7.8. The enzyme cleaves peptide bonds with acidic amino acid residues at the carboxyl side [162].

5.2 Cysteine Proteinases

Cysteine proteinases are sensitive to sulphydryl reagents, such as pCMB, TLCK, iodoacetic acid, iodoacetamide, heavy metals, and are activated by reducing agents such as potassium cyanide or cysteine, dithiothreitol, and EDTA. The occurrence of cysteine proteinases has been reported in only a limited number of fungi [13]. Intracellular enzymes with properties similar to cysteine proteinases have been reported in *Trichosporon* species, *Oidiodendron kalrai* and *Nannizzia fulva*. Extracellular cysteine proteinases have been observed in *Microsporum* species, *Aspergillus oryzae*, *Sporotrichum pulverulentum* [167], and *Bacteroides gingivalis* [168]. Most of these enzymes are active at pH 5–8. Some, though not all, are stimulated by reducing agents [13].

A number of acellular and cellular slime moulds, including *Dictyostelium discoideum* [169, 170], produce cysteine proteinases which exhibit optimum activity at around pH 5. The most frequent detection of cysteine proteinases has been among the protozoa. The occurrence and properties of these enzymes have been described in detail [13]. Most of the protozoan cysteine proteinases are intracellular and are active at low (acid) pH values.

These enzymes can be divided into two groups, according to their side-chain specificities.

5.2.1 Clostripain Proteinase

Clostripain proteinase is derived from the culture filtrates of *Clostridium histolyticum*. The enzyme is specific against basic amino acid residues at the carboxyl side of the cleavage point. The enzyme has a molecular weight of 50 kDa, an isoelectric point of pH 4.8, and is sensitive to TLCK but not TPCK [162].

5.2.2 Streptococcal Proteinase

This enzyme is produced by group A *Streptococcus* species as a zymogen which is autocatalytically converted to the active enzyme. The proteinase has a broad amino acid specificity, a molecular weight of 32 kDa, and an isoelectric point of pH 8.4 [162].

5.3 Aspartic Proteinases

Aspartic proteinases are characterised by maximum activity at low pH (3–4) and insensitivity to inhibitors of the other three groups of enzymes. They are widely distributed in fungi, but are rarely found in bacteria or protozoa. Most aspartic proteinases are sensitive to epoxy and diazo-ketone compounds in the presence of copper cations. They are also inhibited by pepstatin or *Streptomyces* pepsin inhibitor. However, aspartic proteinases of *Aspergillus niger*, *Scytalidium lignicolu* and a number of basidiomycetes belonging to the family *Tricholomataceae* are insensitive to these specific inhibitors [171].

Most aspartic proteinases have molecular weights in the range 30–45 kDa, and their isoelectric points are usually in the range pH 3.4–4.6. These enzymes are specific against aromatic or bulky amino acid residues on both sides of the cleavage point. Catalytic activities involve two aspartic acid residues. The catalytic mechanism of the aspartic proteinases requires the initial binding of a water molecule at the active site before nucleophilic attack on the substrate peptide bond [172]. Many of the fungal aspartic proteinases are unstable above neutral pH and are not found in cultures growing at neutral or alkaline pH [162].

Two types of aspartic proteinases have been recognised; pepsin-like and rennin-like.

5.3.1 Pepsin-Like Proteinases

Pepsin-like aspartic proteinases are widely distributed among the moulds, including *Aspergillus, Penicillium, Rhizopus, Trametes* [13] and *Neurospora crassa* [173]. The enzymes are extracellular and some have been used commercially in processes such as soybean protein hydrolysis in soy sauce manufacture. The microbial pepsin-like proteinases exhibit many similar physical and chemical properties to animal pepsin, including low esterolytic activity [162].

5.3.2 Rennin-Like Proteinases

Rennin-like proteinases share several common properties with the pepsin-like proteinases. However, unlike the pepsin-like enzymes, they are capable of clotting milk in a manner similar to animal rennins. Aspartic proteinases exhibiting rennin-like activity have been isolated from several microorganisms, including *Endothia parasitica*, *Mucor* species and *Aspergillus candidus* [162] (Sect. 6.2.2). The enzymes from *Endothia* and *Mucor* have commercial applications in cheese manufacture (Sect. 6.4).

5.4 Metalloproteinases

These enzymes all have pH optima between pH 5–9 and are sensitive to metal-chelating reagents, such as EDTA, but are unaffected by serine proteinase inhibitors or sulphydryl agents. Many of the EDTA-inhibited enzymes can be reactivated by ions such as zinc, calcium, cobalt. Metalloproteinases are widespread, but only a few have been reported in fungi. Most of the bacterial and fungal metalloproteinases are zinc-containing enzymes, with one atom of zinc per molecule of enzyme [174]. The zinc atom is essential for enzyme activity. Calcium is required to stabilise the protein structure. The amount of calcium in metalloproteinases varies from four atoms per molecule for the *Bacillus thermoproteolyticus* thermolysin to less than 0.2 atoms per molecule for the *Aeromonas proteolytica* enzyme.

Metalloproteinases can be divided into five groups: acid, neutral and alkaline proteinases, and the Myxobacter proteinases I and II.

5.4.1 Acid Metalloproteinases

This group of enzymes was proposed by Gripon et al. [174] to represent enzymes which have lower pH optima (5–6), lower molecular weights (19–20 kDa) and exhibit unique specificity towards synthetic peptides and oxidised insulin B-chain. The proteinases of *Penicillium caseicolum* and *P. roqueforti* and one of two metalloproteinases of *Aspergillus sojae* and *A. oryzae* are included in this group. The *Penicillium* proteinases are also insensitive to phosphoramidon, a specific neutral metalloproteinase inhibitor [13].

5.4.2 Neutral Metalloproteinases

Neutral metalloproteinases show specificity for hydrophobic or bulky amino acid residues. They have pH optima near pH 7 and are in the 30–40 kDa range. Neutral metalloproteinases are widely distributed in microorganisms, especially among the bacilli and *Aspergillus* species. The best known neutral metalloproteinase is probably the thermolysin produced by *B. thermoproteolyticus* [162] (Sect. 6.2.2.1).

5.4.3 Alkaline Metalloproteinases

Alkaline metalloproteinases are produced by organisms such as *Pseudomonas aeruginosa* and *Serratia marcescens*. They exhibit very broad specificity, with optimum activity at pH 7–9, and are slightly larger than the other metalloproteinases (48–60 kDa) [162].

5.4.4 Myxobacter Proteinase I

This enzyme has a molecular weight of 14 kDa and an optimum pH 9, and is specific for small amino acid residues at either side of the cleavage site. This proteinase lyses cell walls of *Arthrobacter crystallopoites* and other Gram positive bacteria. A zinc-containing β-lytic proteinase of *Sorangium* has similar specificity toward the oxidised insulin B-chain [162].

5.4.5 Myxobacter Proteinase II

This enzyme has a molecular weight of 17 kDa and a pH optimum of 8,5–9.0, is specific for lysine residues on the amino side of the cleavage site, but does not lyse bacterial cells. The enzyme is stable from pH 3 to 9 and is stable at 50 °C for 18 h [162].

6 Applied Aspects

6.1 General Applications

As already mentioned in the Introduction, microbial proteinases play an important role in industrial processes, accounting for approximately 40% of the total worldwide enzyme sales. The proteinases of *Bacillus* spp., *Mucor* spp., and *Aspergillus oryzae* account for the bulk of the enzyme production and sales. Even so, proteinases produced by other species are also employed in numerous commercial processes. The processing of cereals, beer, chocolate-cocoa, wines, eggs and egg products, animal feeds, fish, legumes, meat, milk protein hydrolysates, as well as baked goods and cheese requires proteinases as an integral part of manufacture.

Proteinases, and indeed other microbial and non-microbial enzymes, offer several advantages over alternative physical and chemical manipulations in food processing. The main advantage is that they catalyse a specific reaction, avoiding potentially undesirable side reactions which may be caused by less specific processing methods. Also, enzymes are present at low concentrations and, therefore, do not need to be removed after use. Another important advantage offered by enzymes is that they can effect the necessary reactions under suitable conditions of pH and temperature, whereas many chemical or physical treatments require extremes of pH or temperature.

The large success of microbial proteinases in food and other biotechnological systems can be attributed to the broad biochemical diversity of the microorganisms, to the genetic manipulation of the organisms and to improved techniques for enzyme production and purification. Microbial enzymes also conform more closely to the required characteristics, such as cost of using the enzyme, activity at optimal conditions, safety of the enzyme, and availability at required purity and stability, than do animal and plant proteinases [175].

Table 2. Commercial applications of non-microbial proteinases

Proteinase	Application
Calf rennet	Cheese manufacture
Pancreatic	Bating
(Pancreatin,	Food & general hydrolysis
Trypsin,	Analytical
Pepsin)	Pharmaceutical
Papain	Chillproofing
	Meat tenderisation
	Protein hydrolysis
	Animal feeds
	Leather
	Textile
	Pharmaceutical
Malt proteinases	Brewing

In order to obtain high and commercially viable yields of a microbial enzyme, an over-producing strain has to be isolated and its growth and production optimised. The steps taken from screening organisms for enzyme production have been summarised in detail elsewhere [176,177].

Although microbial proteinases are the most important enzymes employed commercially, animal and plant proteinases, especially calf rennet, pancreatic proteinases, malt proteinases, and papain, still play important roles in the food industry (Table 2). These enzymes are unlikely to be displaced by microbial proteinases, except where enzyme supplies become limited, as in the case of calf rennet which has been replaced by fungal proteinases as milk coagulant. In the following sections some of the most important commercial applications of microbial proteinases will be discussed in more detail.

6.2 Industrial Production of Extracellular Enzymes

Commercial production of extracellular enzymes is in principle very simple, involving cultivation of a microorganism and subsequent recovery of the enzyme. The most important aspect of commercial enzyme production is the use of economical and reliable processes which also meet the strict safety and hygiene requirements. This requires the use of an organism which grows on an inexpensive medium and produces constant, high yields of enzyme in a short time. Secondary enzyme activities and content of metabolites must, however, be minimal. Simple and inexpensive recovery of the enzyme leading to a stable product with an acceptable appearance and which can be handled safely is also important, as is the safety of the process to the workers and of the effluents to the environment.

These objectives are fulfilled by the combined optimisation of strain properties and process parameters. Optimisation of strain properties, mainly by the development of suitable mutants, usually offers an inexpensive and permanent solution to the problem.

6.2.1 Major Producers

The major producers of industrial enzymes are Novo Industri A/S, Denmark, Gist-brocades, Netherlands, and Miles Laboratories, U.S., controlling about 70% of the market between them (Table 3). These firms are the major suppliers of proteinases for detergents.

Table 3. Major producers of industrial enzymes

Company	Market Share (%)
Novo Industri A/S (Denmark)	40
Gist-brocades (Netherlands)	20
Miles Laboratories (USA)	10
Hansen (Denmark)	5
Sanofi (France)	5
Finnish Sugar (Finland)	5
Others	15

The major producers of rennin for the dairy industry are the Danish firm Hansen, and Sanofi, a subsidiary of the French oil and chemical company Elf Aquitaine. Each company controls about 5% of the market [178].

6.2.2 Major Microbial Strains

Most commercial enzymes are produced by organisms belonging to the *Bacillus* and *Aspergillus* genera. The majority of the *Bacillus* species are harmless, non-toxin producing saprophytes, which are easy to grow in high density without a requirement for expensive growth factors. Undesirable aspects of *Bacillus* metabolism for enzyme production, such as the formation of heat-resistant endospores, or the production of antibiotics or unwanted enzyme activities, can be eliminated by the development of mutants deficient in these properties. The asporogenic mutants produce improved proteinase yields [179], probably due to a longer production phase. Proteinase-hyperproducing strains of bacilli have also been developed [180, 181].

The *Aspergillus* species, like *Bacillus*, are highly variable and widespread, and the majority of the species are non-pathogenic and non-toxin forming. The most frequently used species for enzyme production are the *A. niger* and *A. oryzae* groups. The *Aspergilli* are usually haploid in their vegetative phase and are therefore easily mutable but their mutants are rather unstable.

6.2.2.1 Bacillus Proteinases

The *Bacillus* proteolytic enzymes are by far the most important group of enzymes produced commercially. Their major application is in the detergent industry, accounting for about 35% of the total microbial enzyme sales. The enzymes are also used in protein, brewing, meat, photographic, leather and dairy industries.

Two different types of alkaline proteinases have been identified and characterised, differing from each other by 58 amino acids [182]. They are Subtilisin Carlsberg, produced by *B. licheniformis*, and Subtilisin Novo, or Bacterial Protease Nagase (BPN′), synthesised by *B. amyloliquefaciens*.

The Carlsberg enzyme was discovered in 1947 by Linderstrom, Lang and Ottesen at the Carlsberg Laboratory. Commercial utilisation of the enzyme began in 1960 when the enzyme was found to have excellent properties for use in detergents. The enzyme is now the most widely used detergent proteinase. Annual production of the enzyme is equivalent to about 500 tons of pure enzyme protein.

The Novo proteinase has been used for industrial enzyme production for over 50 years, and until 1960 nearly all bacterial proteinase preparations were made by *B. amyloliquefaciens*. Currently, Subtilisin Novo is used only to a minor extent.

The Carlsberg and Novo Subtilisins are closely related. The enzymes have a molecular weight of around 27.5 kDa [182], a temperature optimum at 60 °C and a pH optimum at 10. The active site of the two enzymes is formed by the residues Ser (221), His (64) and Asp (32) [182], and the enzyme activity is inhibited by reagents such as DFP and PMSF. No cysteine residues are present in the molecule. Both enzymes have a broad, although different, specificity and hydrolyse most peptide bonds and some ester bonds. The Carlsberg proteinase is also effective in transpeptidation and transesterification. The Novo enzyme, unlike the Carlsberg enzyme, is slightly dependent on Ca^{2+} for stability. The Carlsberg proteinase has slightly broader pH and temperature ranges than the Novo enzyme. The proteinases are rapidly destroyed by oxidising agents such as hypochlorite or hydrogen peroxide. However, enzyme activity is unaffected by H_2O_2 stabilised with sodium perborate.

The most recent detergent proteinases, Esperase® and Savinase™T, produced by alkalophilic *Bacillus* species, were discovered in 1967. The bacilli can grow at pH values above 10 and the proteinases are active and stable up to pH 12. Apart from their better pH stability, they have similar properties to the subtilisin proteinases. They have a broad specificity and exhibit esterase activity. Their molecular weight is in the range 20–30 kDa, and the pI around 11. In detergents the new proteinases are superior to Subtilisin Carlsberg, especially at high alkalinity or if sequestering agents such as citrate or gluconate are used instead of the traditional TPP. The proteinases are also very useful in the dehairing process, although their application has been limited.

The neutral metalloproteinases from bacilli are also produced commercially, but to a much lesser extent. The metalloproteinases, like the serine enzymes, are widely distributed in nature. They are produced by several *Bacillus* species, but only the metalloproteinases produced by *B. amyloliquefaciens* and *B. thermoproteolyticus* are used for industrial purposes. The enzymes require Zn^{2+} for activity and Ca^{2+} for stability, and are inhibited by sequestering agents such as EDTA. They contain no disulphide bridges in their peptide chain. The pH optimum is neutral and they are in general not as stable as the serine enzymes. One exception is the *B. thermoproteolyticus* proteinase, thermolysin, which is stable up to 80 °C. In thermolysin, the zinc atom is located at the bottom of a deep cleft, which separates the two lobes of the molecule, and is bound to two histidine residues and a glutamate residue. The active site consists of 6 amino acids close to the zinc atom. Four calcium atoms, which appear to be important for the heat stability of the enzyme, are bound to the molecule. Thermolysin contains 316 amino acids and has a molecular weight of 34,4 kDa. The *B. amyloliquefaciens* enzyme has similar properties to those of thermolysin, but is a slightly larger molecule, containing 326 amino acids, with a molecular weight of 37 kDa. The enzyme is considerably less heat stable than thermolysin, probably due to the presence of fewer hydrophobic amino acids and only two calcium atoms. The metalloproteinases hydrolyse preferentially peptides with hydrophobic side chains, such as phenylalanine and leucine. They have very weak esterase activity. The metalloproteinases are currently used for leather bating, hydrolysis of barley proteins in breweries, and in the food industry.

6.2.2.2 Aspergillus Proteinases

The most important commercial application of the *Aspergillus* proteinases is in soy sauce production. In this process cultures of *A. oryzae* and *A. sojae* are used to hydrolyse the soybean proteins to amino acids. The organisms release several proteinases and exopeptidases during growth on the proteins.

Commercial preparations from *A. oryzae* contain acid, neutral, and alkaline proteinases, and exhibit activity in the pH range 4–11. The acid proteinase has an optimum pH at 4.0–4.5 and is stable between pH 2.5 and 6.0. The active centre is similar to that of pepsin [123)]. *A. oryzae* also produces two neutral proteinases which are metalloproteinases and are inhibited by sequestering agents. One of the enzymes has optimum activity at pH 7 and is stable in the pH range 5.5–12.0, but is rapidly inactivated above 50 °C. The other neutral proteinase has optimum activity at pH 5.5–6.0 and is relatively thermostable, losing only 30% activity at 90 °C after 10 min [183]. The alkaline proteinase is a serine proteinase with similar properties to the serine proteinases of the *Bacillus* species. The enzyme has optimum activity at pH 7.0–8.5 and is stable between pH 4.5 and 9.0, but is temperature sensitive, being rapidly inactivated at 60 °C. The alkaline proteinase is stabilised by Ca^{2+} and inhibited by serine reagents and by potato inhibitor [184]. Acid proteinases are also produced commercially by black Aspergilli, especially *A. saitoi* (synonymous with *A. phoenicis*) and *A. niger* var *macrosporus*. *A. niger* produces two acid proteinases, with optimum activities at pH 2.0 and pH 2.6, respectively. The *A. saitoi* proteinase, Aspergillopeptidase A, has an optimum activity at pH 2.5–3.0 and a temperature optimum at 30 °C. It is stable between pH 2 and 5. The enzyme hydrolyses preferentially peptide bonds with hydrophobic side chains. The proteinase has very little peptidase, amidase or esterase activity. It activates trypsinogen and chymotrypsinogen. The molecule is a monomer of molecular weight 34–35 kDa and contains two cysteine residues which form a disulphide bridge. *Aspergillus* proteinases are used primarily in digestive aids, where the acid pH optimum and the large number of concomitant enzyme activities (α-amylase, cellulase, glucoamylase and pectinase) are beneficial. The *A. oryzae* proteinases are used extensively in the U.S.A. for flour treatment, where they exert gentle hydrolysis of the gluten, reducing mixing time of the dough and improving bread quality. The proteinases also find limited use in meat or fish protein hydrolysis under acidic conditions.

6.2.2.3 *Mucor Proteinases*

The milk coagulating ability of a *Mucor* acid proteinase (produced by *M. rouxii*) was first reported in 1921 [185]. However, due to its low milk coagulating activity with respect to its proteolytic action, its application was unsuccessful. This finding led to intensive screening of the *Mucor* species, and a thermophilic strain belonging to the species *M. pusillus* with a high milk coagulating to proteolytic ratio was isolated [186]. In 1965, the related species *M. miehei* was found to form a similar enzyme with equally good milk coagulating properties [187]. Both enzymes are now used extensively in cheese manufacture.

Although the two thermophilic species are related and their proteinases belong to the same group of proteinases, with an aspartate residue in the active centre, the composition and action of the two enzymes is distinctly different. The *M. pusillus* enzyme is a monomer of molecular weight 30 kDa. The molecule contains two cysteine residues which apparently form no disulphide bridge. No carbohydrate moiety is present. The *M. miehei* proteinase is a monomer of molecular weight 38 kDa and contains about 6% carbohydrate. Both enzymes preferentially hydrolyse peptide bonds with aromatic and hydrophobic side chains and exhibit optimum activity at pH 4.0–4.5. They are stable in the pH range 3–6. The *M. miehei* proteinase is more heat stable than the *M. pusillus* enzyme. The milk coagulating ability of the two proteinases, especially that from *M. pusillus*, is slightly dependent on the presence of calcium.

6.2.2.4 *Endothia Proteinase*

The acid proteinase from *E. parasitica*, a pathogen of chestnut trees, was discovered in 1963 [188]. The enzyme exhibits good milk coagulating activity and is less dependent on pH variation in milk than is calf rennet. It is, however, temperature sensitive, being inactivated within 5 min at 60 °C.

The proteinase is a monomer of molecular weight 34–39 kDa, with no carbohydrate moiety. The enzyme is stable in the pH range 4.0–5.5 with optimum activity at pH 4.5. The enzyme has a broader specificity and is less dependent on calcium for its milk coagulating activity than calf rennet and the *Mucor* enzymes.

Its use as a milk coagulant has been limited because of its high proteolytic activity. The *Endothia* proteinase is used in the production of Emmenthal cheese, for which it is superior to the *Mucor* proteinases, since the cheese undergoes a high temperature treatment during which the enzyme becomes rapidly inactivated.

6.2.3 The Cultivation Process

Two methods of cultivation are used, namely semisolid and submerged. Semisolid cultivation is the traditional method of microbial enzyme production. It is still used for the production of a number of fungal enzymes, such as the proteinases from *Aspergillus* and *Mucor* species. Submerged culture methods, however, dominate in enzyme production because of lower handling costs and risks of infection, and the easier adaptation of modern methods of control to these processes.

6.2.3.1 Semisolid Cultivation

Microorganisms are grown on moist, solid media, containing preferably wheat bran with various additives. Cultivation is performed either in trays with a substrate thickness of 1–10 cm, or in rotating drums tumbling the substrate to facilitate aeration.

The medium is prepared by mixing bran with water and additives, steam sterilised, then transferred to trays under aseptic conditions. Spores are inoculated either in the autoclave after cooling, or in the cultivation equipment. Aeration is obtained by blowing humidified air over the culture.

Contamination is a major problem in this type of cultivation, as sterilisation of the semisolid medium is difficult and aseptic handling of the sterilised medium is almost impossible. Another disadvantage of this system is the difficulty in controlling medium pH and supplementation with medium components during growth of the organisms. The major advantages of this system, however, are the high aeration rate obtainable and the low water concentration. This leads to a broader variety of enzyme production than would be possible in submerged culture.

Production of microbial rennet by *M. pusillus* is performed in semisolid culture. Suitable strains for rennet production are ATCC 16458 and NRRL 2543. The organism is cultivated on a medium consisting of 60% wheat bran and water for 3 days at 30 °C. The enzyme is then extracted with water. Ammonium salts are added to the bran to increase the yield. During cultivation cellulase, lipase and an unspecific proteinase may also be formed. They can be removed by acid treatment, adsorption on silicon dioxide or other adsorbents. The commercial preparation contains only one proteinase component. (The *M. miehei* proteinase is produced in submerged culture).

Production of high yields of the *Aspergillus* proteinases is also only possible in semisolid culture. Either wheat or rice bran is used, and a high ratio of inorganic nitrogen to carbon is reportedly important for good yields [189]. The best strains for proteinase production are *A. oryzae* NRRL 2160, *A. satoi* ATCC 14332 and *A. niger* ATCC 16513. Concomitant with proteinase release, high concentrations of α-amylase, cellulase, glucoamylase and pectinase are produced. In submerged culture, *A. oryzae* produces only a serine proteinase [190].

The proteinase is usually recovered by extraction of the bran with water and precipitation of the extract with solvent. Unless the preparation is highly purified the final product will contain all the enzymes produced during growth of the organism. Thus, for medical use, the *Aspergillus* proteinases are purified by more refined, laboratory purification methods.

The *Aspergillus* proteinases are marketed in solid form; the *M. pusillus* enzymes are usually sold in liquid form, although solid preparations, made by solvent precipitation or by direct spray-drying of the purified solution, are also used.

6.2.3.2 Submerged Cultivation

Since microbial enzymes are relatively low volume products, the equipment and methods used are often adapted from those used in antibiotic preparation. Only the media and some cultivation conditions vary.

The traditional equipment used for enzyme production are tall cylindrical stainless steel bioreactors with capacities of 10–100 ton, containing strong mechanical agitators and air spargers [177] (Fig. 1).

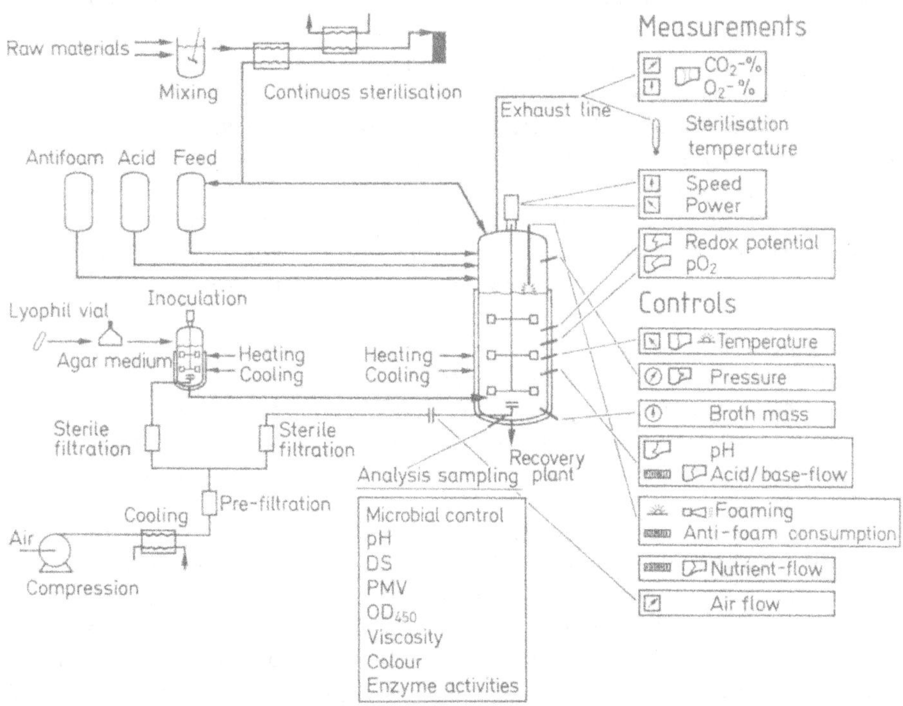

Fig. 1. A typical process for enzyme production. Examples of useful measurements and controls are indicated. (Courtesy of K. Aunstrup, Novo Industri A/S, Denmark)

Table 4. Typical media for proteinase production in submerged culture

Organism	Composition	$g\,l^{-1}$
Bacillus spp.	1. Starch hydrolysate	50
	Soybean meal	20
	Casein	20
	Na_2HPO_4	3.3
	2. Ground barley	100
	Soybean meal	30
	pH 9–10	
Mucor miehei	Starch	40
	Soybean meal	30
	Ground barley	100
	$CaCO_3$	5
Endothia parasitica	Soybean meal	30
	Glucose	10
	$NaNO_3$	3
	Skim milk	10
	KH_2PO_4	0.5
	$MgSO_4 \cdot 7\,H_2O$	0.25

Cultivation media are chosen to be inexpensive (raw materials account for 60–80% of the variable costs of an enzyme cultivation process) and to support good growth of the microorganism. The typical cultivation media used for proteinase production are listed in Table 4.

Compounds such as glucose, which tend to repress enzyme formation, are kept to a minimum during the whole period of cultivation, either by using slowly metabolisable carbohydrates such as starch or lactose, or by adding the compound slowly during growth. The composition of the medium is such that at the end of cultivation total dry substance content and viscosity are low, the cell mass easily separable, and the amount of free carbohydrate and amino acids in the medium minimal. The media are sterilised mainly batchwise in the bioreactor, but continuous heat sterilisation, which is a high-temperature/short-time process, may also be used [177].

Extracellular enzymes are usually produced by batch processes which last 30–150 h. Most processes have a high oxygen demand and require vigorous aeration and agitation. Batch processes are used in preference to continuous systems, because of inefficient medium utilisation during continuous cultivation. The process is often terminated before or after the optimum harvesting time in order to obtain a broth which will facilitate enzyme recovery.

The *B. licheniformis* and *B. amyloliquefaciens* subtilisins are made by cultivation of the organisms on a concentrated medium with a high protein or protein hydrolysate content (Table 4), at neutral pH and 30–40 °C. Carbohydrates are usually added during the process. Production methods of proteinases from the alkalophilic *Bacillus* species are similar to those used for making the subtilisins. Alkali, such as sodium carbonate, or salts of metabolisable acids, such as lactates, are usually added to maintain the medium pH above 7.5 to prevent cell death [191,192].

B. licheniformis also releases α-amylase and α-glucosidase, but the proteinase hydrolyses all proteins in the medium and at the end of the cultivation process it is practically the only protein in the broth. A cell-bound penicillinase, which can be removed from the medium with the cells, is also produced. The strains used for Subtilisin Carlsberg production cannot produce the antibiotic bacitracin, which is a competitive inhibitor of the proteinase, nor D-glutamylpolypeptide, which makes the medium viscous and slimy. Yields of the Subtilisin Carlsberg protein of over 10% of the initial protein content are common. The enzymes are sold mainly as dust-free granulates containing 1–5% active enzyme protein. Liquid preparations of the enzyme contain about 2% proteinase [177].

Similarly, the alkalophilic proteinases are the major enzymes in the cultivation medium, representing over 90% of the enzyme content. The alkalophilic proteinases are sold either as granulates or as dust-free powder containing 2% active enzyme content.

B. amyloliquefaciens, on the other hand, produces several other extracellular enzymes during cultivation, including α-amylase, β-glucanase, neutral proteinase, and hemicellulase. The latter two enzymes are unstable and are usually present in low concentrations in the final Subtilisin Novo preparation. α-Amylase, however, is present in large amounts, and although methods for removal of the amylase have been described [193], the final preparations usually contain the α-amylase and Subtilisin Novo enzymes. The content of Subtilisin Novo is usually less than 1%.

Isolation of maximum yields of the *B. amyloliquefaciens* metalloproteinase requires short cultivation times. The enzyme, which is produced together with the alkaline proteinase, is very unstable and if not removed quickly may be destroyed during cultivation. Thermolysin is prepared by cultivating *B. thermoproteolyticus* at 55 °C for 24 h [194]. If metalloproteinases without interfering enzymes are required, either the broth is fractionated after recovery [195], or mutants free from alkaline proteinase are used [196].

The microbial rennets of *M. miehei* and *E. parasitica* are also produced in submerged culture. *M. miehei* is usually grown for 7 days at 30 °C in a medium of pH 5–7, containing starch, soybean meal and barley (Table 4). Other enzymes such as amylase, cellulase, lipase and esterase are also produced. Lipase and esterase are present in large amounts and usually have to be removed from the proteinase preparation. The *M. miehei* proteinase, like the *M. pusillus* enzyme, is usually marketed as a liquid preparation, although solid preparations are used in some areas.

E. parasitica (e.g. strain ATCC 14729) is grown at 28 °C for 48 h in a medium composed of soybean meal, pH 6–7 (Table 4). Due to instability of the enzyme it has to be recovered quickly at low temperature and, preferably, in the absence of oxygen. The enzyme is marketed as a solid preparation.

6.2.3.4 Enzyme Recovery

The most common methods leading to commercial grade enzyme products are shown in Fig. 2. Upon termination of a batch process the solution is rapidly cooled to about 5 °C, to prevent microbial contamination. Solid materials, such as bacterial cells, fungal mycelium and medium components, are removed by filtration or centrifugation. Separation is aided by pretreating the solution with a coagulating or flocculating agent. It can be precipitated with inorganic salts, such as calcium sulphate or calcium phosphate, by the addition of sodium sulphate. An alternative,

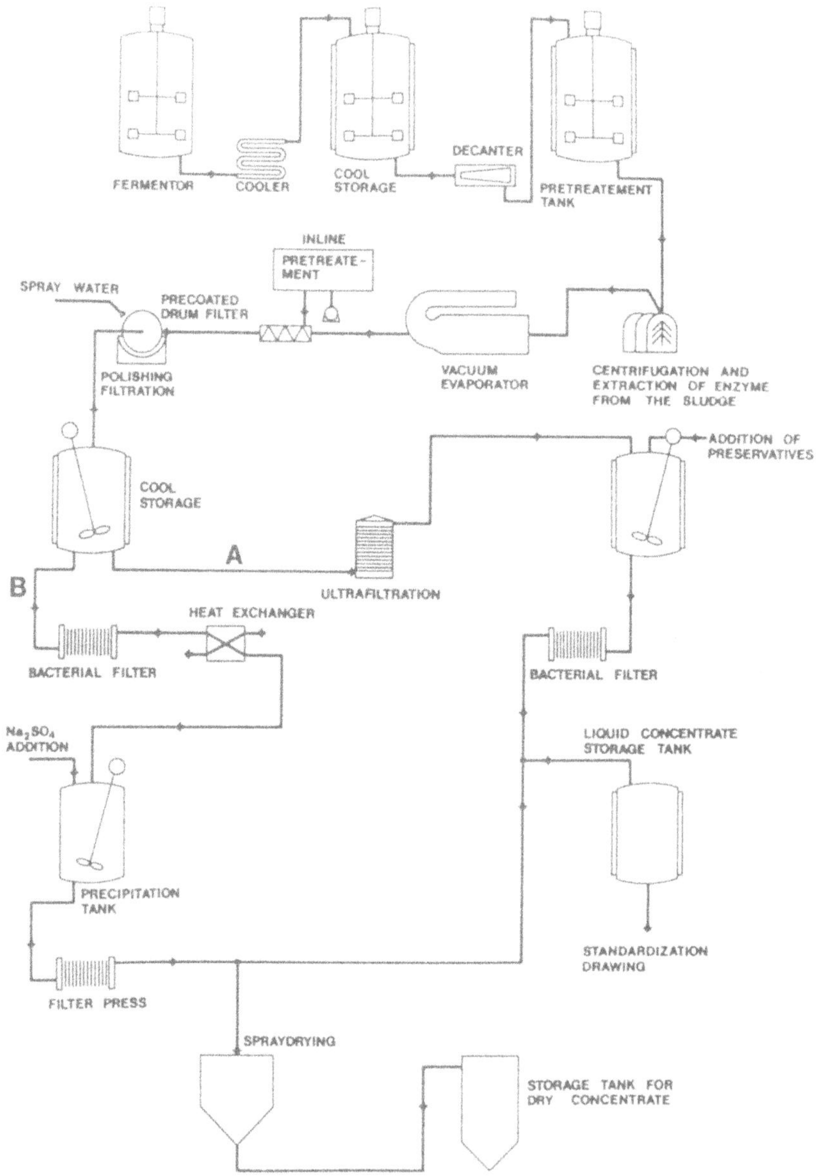

Fig. 2. Outline of an extracellular microbial enzyme recovery process. (Courtesy of K. Aunstrup, Novo Industri A/S, Denmark)

and more efficient, methods is flocculation using synthetic polyelectrolytes such as polyamines. In some cases filter aid, such as diatomaceous earth, must be added before filtration.

In many cases the separation process is aided by letting the spent medium set so that the colloids aggregate before separation. This is, however, limited due to costs and contamination risks. The separation process is performed on vacuum drum or leaf filters, or in high-speed disc centrifuges.

The enzyme is initially concentrated either by evaporation in multistage vacuum evaporators, or by ultrafiltration. The ultrafiltration process is in many cases advantageous, since it is inexpensive, it enables the removal of substances of molecular weight below 10 kDa, and can be performed at a low temperature (5 °C) which minimises loss of activity and risk of contamination. The enzyme solution is then clarified by a polishing filtration method, and the remaining microorganisms are removed by germ filtration on cellulose-containing filter pads. This filtrate is subsequently mixed with stabilisers and preservatives. Stabilisers, such as calcium salts, proteins, starch hydrolysates, and sugar alcohols, are used to increase the storage stability of the enzyme preparation. Preservatives such as sodium chloride (18–20%), or benzoate, parabene, or sorbate, are usually added to liquid enzyme preparations.

Liquid enzyme preparations are preferred to their solid counterparts, as they are cheaper to produce and are safer and more convenient to apply. However, in cases such as for flour treatment or powder detergents, solid preparations have to be used. Solid enzymes are prepared either by direct spray-drying of the enzyme solution or by precipitation. Spray-drying is the simplest method, but is expensive, subjects the enzymes to high temperatures and oxidation conditions, and does not remove impurities. Also, since low molecular weight substances make the product sticky or hygroscopic, only solutions concentrated by ultrafiltration can be used.

Greater purity is obtained when the enzyme is precipitated with solvents such as acetone or ethanol, or organic salts such as ammonium or sodium sulphate. Enzymes used in detergents must be precipitated with sodium sulphate since the ammonium salt is not acceptable. The precipitate formed is removed by filtration and dried. Solvent precipitation yields a product of higher purity and activity than does salt precipitation. However, since solvents are potentially explosive and their recovery is expensive, large scale operations are usually performed with salts. Fractional or multiple precipitation with intermediate purification steps improves the precipitation process.

The purification process sometimes includes a step which removes an undesirable side effect. One example is the removal of lipase in the production of microbial rennet by *M. miehei*. Following cultivation the spent medium is left at a pH below 3.5 for a few hours. The lipase activity present in the spent medium becomes reduced to less than 10%, whereas loss of rennet activity is insignificant.

The dried enzyme preparations form large lumps which are ground to a fine powder in a mill and the activity standardised by the addition of inert compounds, such as salt or lactose. Enzymes used for flour treatment, such as the *Aspergillus* proteinases, are usually standardised with flour.

Until the late 1960's all solid enzymes were sold as dusty powders with a small particle size. Problems in handling of the powders, especially in the detergent industry, led to the development of dust-free granulates. An additional advantage of the granulates is improved storage stability.

Enzymes for the detergent industry are usually granulated by embedding the enzyme into spheres of a waxy material consisting of a nonionic surfactant by means of a spray-cooling or prilling process. Alternatively, the enzyme is mixed with a filler,

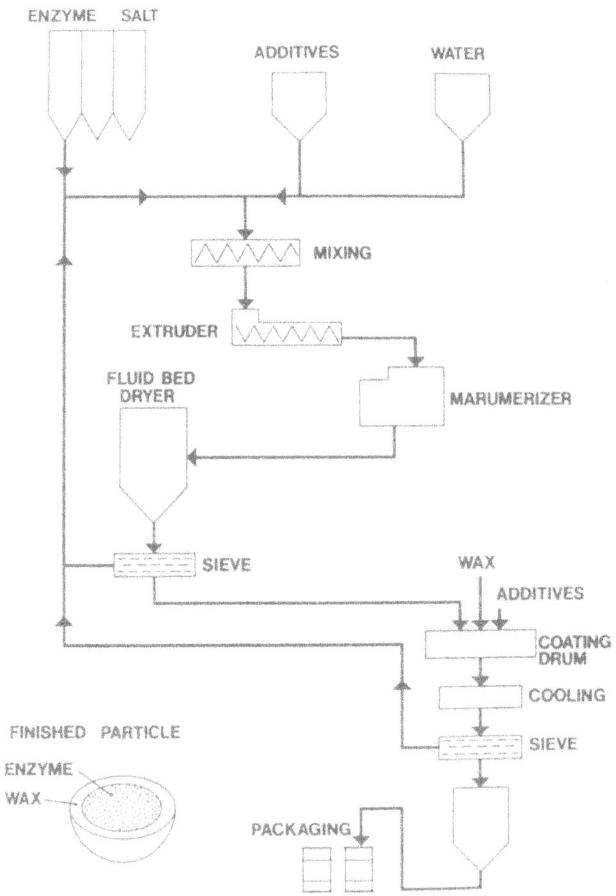

Fig. 3. Preparation of dust-free enzymes. (Courtesy of K. Aunstrup, Novo Industri A/S, Denmark)

a binder, and water, then extruded, and subsequently formed into spheres in a so-called marumeriser. The spheres are then dried and coated with a layer of inert material such as wax or a soluble cellulose derivative [177] (Fig. 3). The commercial granulates contain 1–5% active proteinase protein.

6.3 Detergents

6.3.1 Introduction

Proteinases were first used in detergents as early as 1913 when Röhm and Haas from Germany introduced the pancreatic enzymes in a detergent called „Burnus®". However, it was not until the 1960's that proteinases were effectively incorporated into detergent powder. In 1960 an alkaline proteinase from *B. licheniformis* named Subtilisin Carlsberg (Alcalase®) was developed by Novo Industri A/S, Denmark, and marketed in a detergent under the trade name "Biotex®". By 1969 the market share in Europe and the United States for enzyme detergents was as high as 50%. Soon after 1969 the powdered enzyme detergents received unfavourable publicity when some workers handling the enzyme, prepared as a finely ground powder, developed allergic reactions and a slump in sales resulted. This led to the development of new enzyme applications and the introduction of dust-free, encapsulated products. Following a favourable report by the Ad Hoc Committee on Enzyme Detergents saying that enzyme-containing powders were not hazardous, sales of enzyme washing powders once again increased. The current European market share of proteinase containing detergents is estimated at approximately 75% [197]. In Japan and USA enzyme detergents occupy about 55% and 30% of the market, respectively [198]. Detergent enzymes account for about 25% of the total world-wide enzyme sales [1,178]. The US market for detergent products was worth $ 25.3 billion in 1986 and is expected to rise to $ 30 billion in 1995 [199].

6.3.2 Detergent Components

The pH of laundry detergents is generally 9–10.5 and temperature requirements can be as high as 95 °C. Detergents also contain oxidising and chelating agents which can hinder the activity of certain enzymes. Some surface-active agents, especially cationic surfactants, also denature enzymes. Non-ionic surfactants, however, usually

Table 5. Formulation of various types of detergents

Soaking (pre-wash) detergent	
Anionics (LAS)	6–15%
Soaps	0–5%
Nonionics	1–3%
Sodium TPP	10–20%
Zeolite	10–20%
Sodium silicate	0–5%
Sodium CMC	0.5–1.0%
Optical brighteners, perfume	0.1–0.2%
Proteinase (e.g. Alcalase 2.0 T or Savinase 4.0 T)	0.4–0.8%
Amylase (e.g. Termamyl 60 T)	0.4–0.8%
Sodium sulphate	to 100%
pH	8.0–9.5
Liquid detergent	
Anionics	5–30%
Soaps	0–15%
Nonionics	10–40%
Solubilisers, solvents, stabilisers	5–20%
Optical brighteners, perfume	0.1–0.5%
Proteinase (e.g. Alcalase 2.5 L/SL, Esparase 8.0 L/SL, or Savinase 8.0 L/SL)	0.4–0.8%
Amylase (e.g. Termamyl 300 L, Type D)	0.2–0.4%
Water	30–50%
pH	7.0–9.5

Table 5. (continued)

Heavy duty detergent	
Anionics (LAS)	7–15%
Soaps	2–4%
Nonionics	1–5%
Sodium TPP	10–20%
Zeolite	10–20%
Sodium perborate	15–30%
Sodium silicate	4–8%
Sodium CMC	0.5–1.0%
Optical brighteners, perfume	0.4–0,5%
Proteinase (e.g. Alcalase 2.0 T, Esperase 4.0 T, or Savinase 4.0 T)	0.4–0.8%
Amylase (e.g. Termamyl 60 T)	0.4–0.8%
Sodium sulphate	to 100%
pH	9.5–10.5

Dishwashing detergent	
Sodium TPP	20–50%
Sodium metasilicate	10–30%
Sodium bicarbonate	40–60%
Surfactant	3–10%
Proteinase (e.g. Alcalase 2.0 T	1–3%
or Esparase 4.0 T)	1–3%
Amylase (e.g. Termamyl 60 T)	1–3%
pH	9.0–9,5

have no influence on the stability of the enzymes. Hence, the ideal detergent enzyme must be stable at pH 9–10.5, have good temperature stability, be compatible with oxidising and chelating agents, and be effective for stain removal at low enzyme level (0.4–0.8%) in the detergent solution. The *Bacillus* alkaline proteinases have met these requirements successfully. Most of the other proteolytic enzymes are not compatible with the major chemical constituents of detergents. Table 5 presents some examples of enzyme detergent compositions.

The surfactants used are biodegradable, organic compounds, which contain both hydrophobic hydrocarbon and hydrophilic ionic or polar groups. Surfactants reduce the surface tension of water to help in wetting clothing and in emulsifying and dispensing soil. The major surfactants are linear alkylbenzene sulphonates (LAS), alcohol sulphonates (AS), linear alcohol ethoxylates (AE), and alcohol ether sulphonates (AES). AE is effective in cold water systems and also removes body oils from synthetic fabrics. LAS is the preferred surfactant in powder detergents as it is easier to process into powders and is a good foam producer. The main surfactants used in dishwashing detergents are LAS and AES. In Japan, laundry detergents contain AES, while dishwashing detergents are based on AES and alkylamine oxide surfactants.

Perfumes and fragrances are usually added to the product to increase their aesthetics. Most commonly used are alcohols, esters, and aldehydes. CMC is an important ingredient in some heavy duty laundry detergents as it prevents the redeposition of soil onto fabrics.

Enzymes, together with other heat-labile components such as sodium perborate and perfume, are added to the otherwise finished detergent, thus avoiding exposition to high temperature and humidity. Mixing of the enzyme with the detergent components is performed either in a micro-dosing system or by a step-wise procedure [200]. The particle size of the enzyme granules is similar to that of the detergent components. This ensures a homogeneous blend and prevents segregation of enzyme particles during transport and handling of the detergent.

6.3.3 Detergent Proteinases

Proteinases are valuable detergent components in most washing processes. The enzymes remove not only the obvious stains, such as blood, but also other less obvious material including proteins from body secretions and skin particles, and food such as milk, egg, meat and fish. In the absence of proteinases, proteinaceous dirt coagulates on the fabric as a result of the washing conditions. The high temperatures and pH, and the action of the surfactant and sequestering agents, used in washing processes dissolve or disperse most of the dirt components, and the bleaching agent decomposes the undissolved dye. The process, however, causes the protein material to precipitate onto the fabric. The precipitation occurs mainly at about 50 °C when the oxidising agent, hydrogen peroxide, is released in the perborate-containing detergents. The coagulated protein also retains other dirt components, such as soil, lipids, and carbohydrates, and is subsequently difficult to remove. Failure to remove the proteinaceous dirt results in a grey and unclean appearance of the fabric after several washings.

The value of proteinases in removing various proteinaceous stains has been proven by numerous tests involving both ordinary clothing and standard test patches soiled with defined compositions of proteins. One such standard patch is the EMPA 116 patch [201], which is cotton fabric impregnated with blood, milk and Indian ink. The test is performed either in standard washing model equipment, such as the

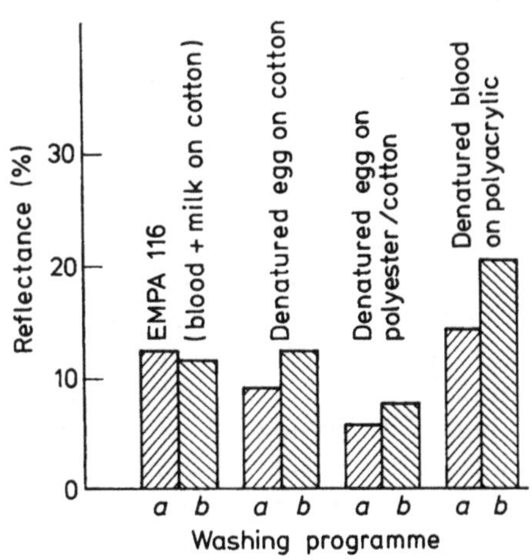

Fig. 4. The effect of Alcalase on different soilings. Test soilings were washed with a commercial USA detergent containing 0.67% Alcalase 2.0 T [200] at (a) pH 9.2 and 20 °C for 30 min, or (b) pH 10.0 and 50 °C for 15 min

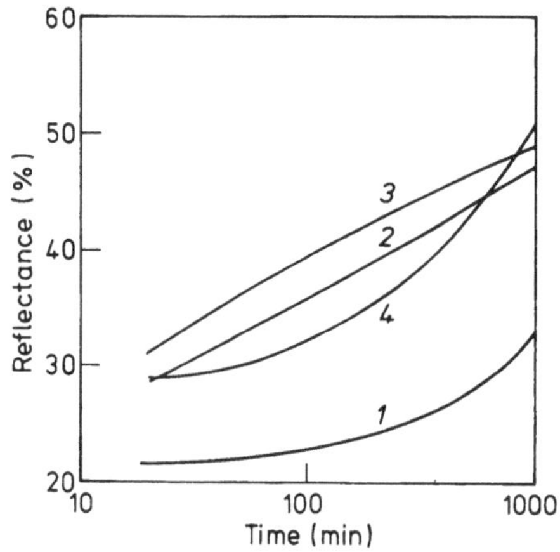

Fig. 5. The effect of pre-soaking on detergent performance. EMPA 116 was soaked with a USA detergent containing 1% proteinase at 20 °C and pH 9.6–9.8, then washed for 20 min [200]. (1) no enzyme, (2) Alcalase 2.0 T, (3) Savinase 4.0 T, and (4) Esperase 4.0 T

Terg-0-Meter or the Launder-0-Meter, or together with ordinary clothing in a household washing machine. A quantitative estimate of the washing efficiency of the enzyme is made by measuring the light reflectance of the washed and dried test patch in suitable equipment. These tests provide valuable information about the efficiency of proteinases in removing different stains and with different washing conditions and detergent compositions [200] (Fig. 4). The ultimate test is always washing of ordinary clothing. Such tests have shown the enzymes to be valuable detergent components in most washing processes.

The most efficient application of proteinases is in a pre-soaking process. The clothes are soaked in the detergent solution at low temperature for a considerable time before the final washing. This enables the enzymes time to work (Fig. 5). In a common European washing process the clothes are prewashed at 20–30 °C for 10–20 min; the temperature is then raised during the main wash to 40, 60 or 95 °C in 30–60 min. The enzyme exerts its function during the prewash and main wash up to about 60 °C. In the USA, water at 50 °C is usually added to the clothing and washing is performed while the suds cool for 10–15 min.

Proteolytic enzymes are also used as spotting agents by dry cleaners. Proteinaceous materials become degraded by the action of the proteinases to smaller molecules, which can subsequently be removed by the dry-cleaning solvents. Proteinases are also added to dishwasher powder detergents [202].

6.3.4 Latest Developments

The success of the *Bacillus* proteinases in detergents initiated a search for alkaline proteinases with improved stability under washing conditions. This led to the discovery and development of a new series of proteolytic enzymes from alkalophilic *Bacillus* species. The enzymes offered a number of technical advantages over the *B. licheniformis* enzymes, namely greater alkali tolerance, better stability in the presence of certain sequestering agents, and higher affinity towards proteinaceous dirt. These improved properties can be attributed to their higher ionic charge (pI 11 c.f. pI 9.4 for Alcalase). Two of the serine proteinases from the alkalophilic bacilli, Esperase® and Savinase™T (both Novo), along with Alcalase, are currently used in detergents [202]. The proteinases from alkalophilic bacilli are especially advantageous for use in heavy duty and liquid detergents.

Sequestering agents are added to detergents to complex calcium and magnesium (i.e. water hardness ions) which may precipitate onto or hold soil to clothes, to buffer pH changes, and to aid oil emulsification by increasing pH. The traditional sequestering agent used in detergents is TPP. However, the risk of phosphate pollution has led to the replacement of TPP with other sequestrants, such as zeolites (e.g. sodium aluminosilicate), citrates, carbonates or nitrilotriacetic acid. The washing properties of these agents are usually not as good as those of TPP, but in their presence the use of enzymes improves the washing efficiency.

Fabric softeners and antistatic agents are increasingly used in the USA laundry detergents. They are cationic surfactants. In powder detergents, distearyldimethylammonium chloride is used since it does not cause the components to cake. In liquids, long chain unsaturated cationic softeners, derived from oleyl alcohol, or a mixture of fatty alcohols from beef tallow, are preferred [199].

The addition of proteolytic enzymes to detergents has increased even more with the current trend towards lower washing temperatures. Currently, the most frequently used washing procedures in Europe are at 40 °C and 60 °C, and temperatures as low as 10 °C can be used [202]. The main reasons for this trend are the possibilities of saving energy, and the greater usage of synthetic fibres which cannot be washed at higher temperatures. The enhanced usage of proteinases compensates for the lower efficacy at lower temperatures. The decrease in washing performance at lower temperatures has also enhanced the demand for other detergent enzymes such as amylases [203], lipases and cellulases [198]. Amylase aids the removal of starch-containing stains by hydrolysing the α-1,4-glycosidic linkages in starch, leading to the formation of soluble dextrins and oligosaccharides. The lipases enhance the removal of fatty stains, which are mainly hydrophobic triglycerides, by degradation to more hydrophilic compounds. However, problems of compatibility with other components, characterisation of fatty soil on clothes, and lipase production at relatively low cost still have to be solved. The latest addition to the detergent enzymes is a special multifunctional cellulase, Celluzyme™ (Novo) [198, 204], produced by the fungus *Humicola insolens*. The enzyme exhibits softening, colour brightening and soil removal properties. It is also compatible with normal detergent formulations. The cellulases are believed to exert their action by removing microfibrils from cotton fibres which develop during wash and wear and which are partly responsible for the adherence of various stains. Removal of the microfibrils restores the cotton fibres to their original structure [198]. The enzyme should be introduced within the next two years.

Since perborate is only effective as a bleaching agent above 60 °C, alternative bleaching agents and bleach activators have been developed. In Europe, bleach activators, such as TAED, pentaacetylglucose or tetraacetylglycouril, are now incorporated into the detergents. The activators increase considerably the bleaching activity of perborate at low temperature [205], by reacting with hydroperoxide ions to yield peracetate ions, which are more active at the lower temperatures. The hydroperoxide ions, formed upon hydrolysis of the perborate ions during washing, do not react fast enough at lower temperatures. In the USA, hypochlorites are the preferred bleaching agents. They are, however, not as compatible with enzymes and fluorescent whitening agents.

Fluorescent whitening agents (also called brighteners or optical bleaches), which are dyes such as diaminostilbenes, are often added to detergents to cancel out the yellowing of fabrics caused by repeated washing. The agents whiten white fabrics and brighten coloured ones. They work by absorbing ultraviolet light and by fluorescing at the blue end of the visible spectrum [199].

The low temperature washing procedures and the introduction of phosphate regulations have led to the development of liquid detergents, especially in the USA where their market share for laundry detergents has increased from 25% in 1985 [203] to 30% in 1986 [199]. Liquid detergents offer other advantages over powders in that they are more easily and accurately measured, are more readily dissolved even in cold water, form no irritating dust, and do not cake on storage.

The premature flowing out of liquid detergents from the powder compartments of washing machines has been stopped by making viscous concentrates, or by using a thixotropic additive, probably bentonite clay [199]. Thixotropy is a rheological property of flowing under stress but gelling in the absence of stress.

One of the recent applications of liquid detergents has been in pre-spotting. Prior to the washing procedure particularly dirty patches on fabrics, such as shirt collars and cuffs, are moistened with the undiluted liquid detergent. The high concentrations of detergent components exert a strong effect on the stains and enable them to be removed during washing.

For use in liquid detergents the enzymes are supplied either as slurries, where the enzyme powder is suspended in a non-ionic surfactant, or as true liquids. The storage stability of the enzyme slurries may be improved by protection with silicates [206]. Proteinases are used in about 50% of the US liquid laundry detergents. By 1990, all liquid detergents in the USA should contain enzymes [207]. Similarly, the USA market share of enzyme-containing powder detergents is expected to increase from the current position of 30% to about 60% by 1990 [207].

6.4 Cheese Manufacture

6.4.1 Introduction

The origins of cheese manufacture are believed to date back many centuries to the time of nomadic communities. The milk was carried in sacs made from animal skins or vessels such as stomachs or bladders. If kept warm the milk became sour and separated into curd and whey. The whey could be drained away and the curd dried to form a firm, cheesy mass. This was an important method of preserving much of the food value of milk and is still used in Arabia in the formation of the dried curd, *Kishk*.

The coagulation of milk was found to be caused by a secretion from the stomach of young ruminants. This led to the use of rennet, an extracted enzymatic secretion of the fourth stomach of calves, as a milk coagulant. Rennet is a partially purified enzyme extract consisting of the major milk-clotting enzyme in the extract, chymosin (rennin), and bovine pepsin, which contributes a small portion of the total milk-clotting activity. Both enzymes assist in the proteolysis of casein during cheese ripening.

6.4.2 Cheese Types

Over the years, as a result of experience and local demands, different ways of manufacturing cheese have been developed in different countries and in different areas within a country. There are currently over 400 varieties of cheese, although only about 18 distinctly different types. The various types of cheeses can be divided into four main classes, according to their moisture content and composition [208].

The variety of cheese produced depends on the type of milk used, on the preparation of the young curds, and on the presence of certain microorganisms [209,210]. Most cheeses are made from cow's milk, but milk from sheep, goats and buffaloes is also used in the manufacture of certain cheese varieties. Casein constitutes about 80% of the total nitrogenous material found in milk (Table 6) and is present mainly in micelles (colloidal-sized aggregates of molecules). The k-casein occurs mainly on the outer surface of the micelle, stabilising it and keeping it in suspension [211].

Table 6. Composition of the nitrogenous substances found in cow's milk

Protein nitrogen (3.40% (w/v))				Non-protein nitrogen (0.16%)
Casein	(2.78%)	Whey proteins	(0.62%)	Amino acids
α-casein	1.67%	β-lactalbumin	0.30%	Peptides
β-casein	0.62%	α-lactalbumin	0.12%	Urea
γ-casein	0.12%	Immunoglobulins	0.07%	Uric acid
k-casein	0.37%	Serum albumin	0,04%	Creatine
		Others	0.09%	

The milk used is cheese manufacture contains a relatively broad range of microorganisms and microbial enzymes, such as proteinases and lipases, which survive the pasteurisation process. In addition, numerous adventitious microorganisms enter the milk during handling in the creamery. Some of these organisms, and the enzymes, play a significant role in the development of flavours of certain cheese varieties.

The manufacture of cheese can be divided into five phases;

1) control of milk properties;
2) coagulation of milk;
3) separation of the solid curd from fluid whey;
4) formation of curd into its final physical form; and,
5) controlled storage of cheese to obtain or maintain the required flavour of the cheese.

The fundamental processes in cheese manufacture are described in detail elsewhere [212].

6.4.3 Starter Cultures

Despite the relatively large milk microflora, starter bacteria are added to milk for cheese manufacture. They are lactic acid bacteria, such as streptococci, leuconostocs, and lactobacilli. Selected species of these organisms are added either as mixtures or as single strain cultures. Mesophilic starters are used in the production of a wide variety of cheese. Thermophilic starters, however, are preferred in the production of cooked cheese varieties, where temperatures of about 45 °C are used.

Starters are added to convert the lactose of milk to lactic acid. This imparts a fresh, acid flavour to curd cheeses and assists in the formation of the rennet coagulum. In addition, they promote characteristic texture formation by causing shrinkage of the curd and moisture expulsion. The organisms also produce traces of flavourful aroma compounds and their proteolytic activity helps in cheese maturation.

The addition of starter cultures to provide a rapid and reliable metabolism of lactose prior to coagulation has replaced the traditional method of dependence on natural contamination of milk with lactic acid bacteria.

6.4.4 Milk Coagulation

The increasing demands for cheese and the declining supplies and higher costs of calf rennet have resulted in the development of microbial rennet substitutes, most notably from *Mucor miehei*, *Mucor pusillus* var. *Lindt* and *Endothia parasitica* [213]. The milk coagulants from these strains account for nearly all of the microbial rennets. In 1981, the microbial coagulants were used in more than one third of all cheeses produced worldwide [1]. The three microbial enzymes exhibit properties similar to those of calf rennet, although they tend to liberate more non-protein nitrogen and have higher proteolytic activity than does calf rennet [214]. However, the flavour, texture and appearance of cheeses made with microbial coagulants compare favourably with cheeses made with calf rennet.

The coagulation of milk by rennet takes place in three phases [211];

1) primary or enzymatic,
2) secondary or clotting, and
3) tertiary or proteolytic phase.

The primary and secondary phases involve milk-clotting reactions whereby the casein fraction of milk becomes converted from a colloidal suspension to a fibrous network. Rennets attack k-casein, liberating macropeptides, which destroys the stability effect of k-casein, leaving an insoluble fraction, called paracasein. The paracasein constitutes the cheese matrix and represents about 99% of the proteins in most cheeses [215].

These two phases can be summarised as follows: —

$$\text{k-casein} \xrightarrow{\text{rennet}} \text{para-k-casein} + \text{macropeptide}$$

$$\text{para-k-casein} \xrightarrow[\text{pH 6}]{\text{Ca}^{2+}} \text{dicalcium para-k-casein}$$

The tertiary or proteolytic phase is a complex phenomenon which is responsible for the ripening of cheese, and involves a diversity of proteinases, as well as lactic cultures and adventitious bacteria.

Curd formation with rennet occurs in about 15–30 min. Rennet action is affected by factors such as amount of milk coagulant, temperature, pH and calcium content of milk, the soluble protein content, especially β-lactoglobulin, and the size of the casein micelles.

The coagulated curd is cut, cooked, salted pressed and finally ripened. Whey proteins are not precipitated during the coagulation period and are released in the whey when the curd is cut.

6.4.5 Ripening

Ripening involves changes in the chemical and physical properties of cheese. The process, which takes place in temperature controlled rooms at 2–16 °C, requires up to four years, depending on the variety of cheese. During ripening, microorganisms and enzymes in the curd hydrolyse fat, protein and other compounds present, leading to the development of characteristic flavours of the cheeses. These changes have been well documented [216].

During ripening, cheese proteins become cleaved at various sites and the protein network, which forms the structural component of fresh cheese curd, loses part of its original structure and the rheological properties of cheese become altered. The action of proteinases during cheese ripening contributes to the great variety of cheeses produced.

The role and mode of action of the various proteinases involved in the degradation of the cheese proteins during cheese ripening are now well defined [217]. The α_{s1}-casein is the first fraction to be hydrolysed, mostly by the action of chymosin [213,217]. The native milk proteinases also play an important role in the primary degradation of cheese proteins [215], especially in varieties such as the Swiss-type cheese where rennet is believed to be inactive [218]. The β-casein of cheese, which maintains the rigidity of curd [219], is hydrolysed either by rennet to β-type peptides, or by the alkaline milk proteinase, plasmin, to γ-caseins. The presence of γ-caseins in almost all cheese varieties indicates that plasmin plays the major role in β-casein hydrolysis [220]. With the exception of the mould-containing cheeses, such as Camembert cheese, where fungi hydrolyse the β-casein in the surface of cheese [221], a large amount of β-casein remains unattacked by proteinases at the end of ripening [222]. The primary phase of proteolysis results in the hydrolysis of cheese proteins to polypeptides [213,217]; the secondary phase involves degradation of the polypeptides to lower molecular weight peptides and amino acids by the action of proteinases of the organisms growing in cheeses [212,223]. The amount of proteolysis depends on the cheese variety. Thus, soft cheeses are proteolysed to low molecular weight peptides and amino acids, whereas in hard cheeses only about 30% of the protein is solubilised [216].

6.4.6 Latest Developments

Since the ripening process is long and expensive, methods for its acceleration have been developed. These include the use of lactose hydrolysed milk, an increase in the number of starter cultures, and the use of added enzymes, including proteinases, lipases and esterases [1,224]. The enzymes are added either in a solid form to the curd after the whey has been drained away, or to the milk earlier in the production process. Both methods are, however, inefficient. The enzyme added to the curd fails to reach the interior and a large proportion of it is not used. Similarly, a large proportion of the enzyme added to milk is lost on removal of the whey. The latter method also alters the subtle flavour of the cheese. These problems may be overcome by the microencapsulation of the enzymes in liposomes [225]. The method, investigated at the Agriculture and Food Research Council's Institute of Food Research in Reading, England, halves the ripening time of cheese. Encapsulation of the enzymes needed to ripen cheese (an exo- and an endopeptidase) enables the enzymes to be evenly distributed in the milk and curd. Enzyme losses in the whey can be reduced from 95% to less than 10% [225].

Multiple enzyme systems (proteinases and lipases) are also used to modify natural cheeses by accelerating many biochemical changes which occur during traditional cheese ripening. The products, called enzyme modified cheeses, are used in formulations such as processed cheese, cheese spreads, and cheese dips. They either enhance the cheese flavour intensity of a formulation without increasing the total cheese solids, or maintain cheese flavour when total cheese solids are reduced.

A new thermal method to measure progress of curd development has also been devised [226] which should result in a more uniform industrial cheese production. Immobilisation of microbial rennets may also help improve cheese manufacture, with the immobilised form being used in continuous clotting processes for cheese production. The immobilised enzyme could also be removed from the product, overcoming the problems encountered with the undesirable proteolytic action of the rennets [227]. Alternatively, recombinant DNA technology can be used to produce calf rennet in *Escherichia coli* cells increasing the availability of the animal coagulant [228].

6.5 Soy Products

6.5.1 Soy Sauce Production

Fungal proteinases have been used for centuries in the Orient for the preparation of soy sauce and other soy products [229]. Soy sauce is a dark brown liquid of distinct fragrance and with a salty taste which is used as a seasoning agent in the preparation of food and as a table condiment [230]. It is prepared by incubating soybeans, wheat and salt with a mixture of mould, yeast and bacteria, and allowing the proteins, carbohydrates and other constituents of soybeans and wheat to be hydrolysed to peptides, amino acids, sugars and other low molecular weight compounds by the enzymes of the microorganisms. Salt is added to prevent the development of toxins from the microorganisms.

Two major kinds of *shoyu*, or soy sauce, are produced in the Orient; the Chinese- and the Japanese-type. The Chinese soy sauce is made from soybeans alone or from a mixture of soybeans and wheat containing a higher percentage of soybeans. In Japan, equal amounts of soybeans and wheat are used. Although soy sauce production originated in China, today Japan is the major *shoyu* producer. *Shoyu* is now sold worldwide, and over one million m^3 of *shoyu* is produced annually [229].

Traditionally, whole soybeans were used in soy sauce manufacture. Presently, most of the soy sauce is produced from defatted soybean flakes. The beans are moistened and steamed under pressure, mixed with roasted and coarsely crushed wheat kernels, and finally inoculated with the starter culture [229, 231]. In Japan, the starters used are strains of *A. oryzae* and *A. sojae*, especially selected for high enzyme activity and the ability to grow rapidly in very thick substrates. The mixture is incubated for 2–3 days at 30 °C and 40% moisture and the resulting mass, called *koji*, is placed in brine.

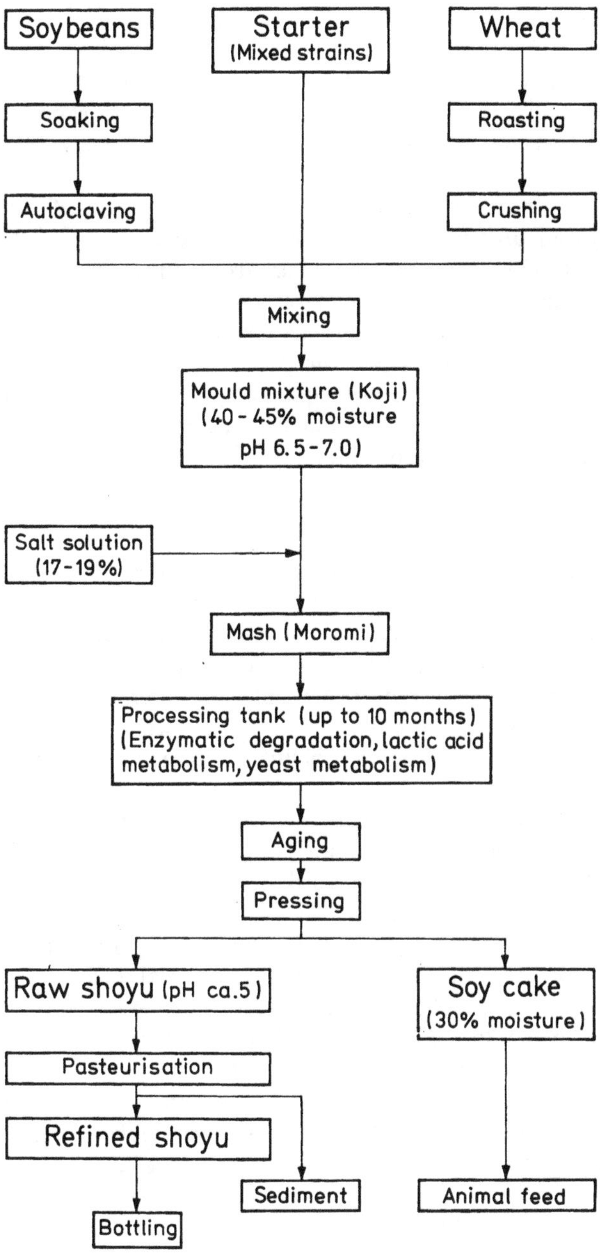

Fig. 6. A basic method for soy sauce manufacture

This mixture, called *moromi* mash, is incubated with yeast and lactobacilli for up to 10 months at a controlled temperature allowing the large molecular weight constituents of the mash to be hydrolysed by the microbial enzymes. The matured *moromi* mash is squeezed through layers of cloth under high pressure to leave a clear solution called raw *shoyu* (unpasteurised soy sauce). The raw soy sauce is pasteurised at

70–80 °C, filtered to remove precipitates, and bottled for market [229, 230, 232] (Fig. 6).

Soy sauce can also be prepared by acid hydrolysis of the bean and wheat mixtures [230]. However, the chemically hydrolysed product lacks the rich flavour of true soy sauce. A large proportion of the soy sauce sold in the United States and Europe is of the chemical variety.

The neutral and alkaline proteinases of the *Aspergillus* species play a significant role in the digestion of soybean protein during soy sauce processing [230, 233, 234]. About 80% of the proteinase produced by the fungi is an alkaline proteinase with a pH optimum of 9–10. In addition, three acid proteinases with optimum pH 3, two neutral proteinases (optimum pH 6–7) and a semi-alkaline proteinase are produced [229, 235]. In Thailand, the addition of *Aspergillus flavus* var. *columnaris* as a controlled inoculum for *koji* has resulted in the production of soy sauce of superior quality to that previously formed [236], the neutral and alkaline proteinases of the *A. flavus* playing a significant role in the improved quality [234].

6.5.2 Other Products

The soybeans used in soy sauce preparation are of high nutritional value, containing good quality protein and a high percentage of essential fatty acids. Consequently, soybeans have been used extensively as raw material for traditional foods such as *tofu* (bean curd), *sufu* (bean cake or Chinese cheese), *miso* (bean paste), as well as *shoyu* [232]. *Miso* is a paste similar to peanut butter. A number of different kinds of *miso* exist. *Miso* is normally used as a soup base, but it is also used as a flavour additive for foods such as fish, meat, vegetable and shellfish. *Miso* is prepared in a similar way to *shoyu* [231, 232].

The formation of *tofu* from soymilk involves mainly coagulation with chemicals such as calcium or magnesium sulphate or acetic acid. The curd is then pressed to remove whey, leaving a soft, but firm, cakelike curd (*tofu*) [232]. Efficient curd formation from soymilk can also be achieved with the plant proteinase, bromelain [237, 238], and many soil bacteria [239]. Attempts are being made to incorporate the use of microbial proteinases in soymilk curd formation on a commercial scale.

Sufu is a soft cream cheese-type product modified by fungi from *tofu*. The *tofu* used for *sufu* production is at first cut into cubes, sterilised at 100 °C for 15 min and finally inoculated with the fungus, *Actinomucor elegans*. The fungus has high proteolytic and lipolytic activity, which act upon the protein and lipid substrates in *tofu*. *Sufu* is usually eaten directly, but may also be cooked with vegetables or meat.

Soy protein may also be added to low pH foods such as beverages, where the problems of gelation, turbidity and off-flavour are encountered. Increased amounts of soy protein are also added to processed meat such as burgers and frankfurters [240] in an attempt to reduce the high costs of meat.

6.5.3 Modification of Functional Properties

Despite the broad application of soybeans in foods, attempts have been made [237, 238] to improve functional properties of the soybean protein and to eliminate shortcomings such as beany flavour [241] and water absorption and retention in gel used for cheese-like foods [240, 241]. The functional properties of the soy protein can be improved by proteolytic modification. The functional properties of the hydrolysate are affected by the method of protein extraction. These properties can be improved by using a membrane isolation method instead of the traditional method [1] to extract the soy proteins. The traditional method of acid extraction produces greater protein degradation and hence greater susceptibility to hydrolysis. The bitterness of the soybean protein hydrolysates increases with increased hydrolysis. At the same time the yield of soluble hydrolysate increases with hydrolysis. Hence, a balance has to be maintained between the amount of hydrolysis and the number of bitter-tasting peptides, although the bitter flavour may be removed by plastein formation [242,245] (Sect. 6.9). Soluble hydrolysates with high solubility, good protein yield and low bitterness can be produced by treating the soy protein with Alcalase at pH 8 and 50 °C. The reaction is stopped by acid precipitation when about 10% of the protein is hydrolysed, leaving soluble protein in the supernatant fraction [246] (Fig. 7). The resulting hydrolysate is used in protein-fortified soft drinks and in the construction of special dietetic feeds, especially for cancer patients with affected eating patterns. The soy hydrolysate can be used in calf milk replacers.

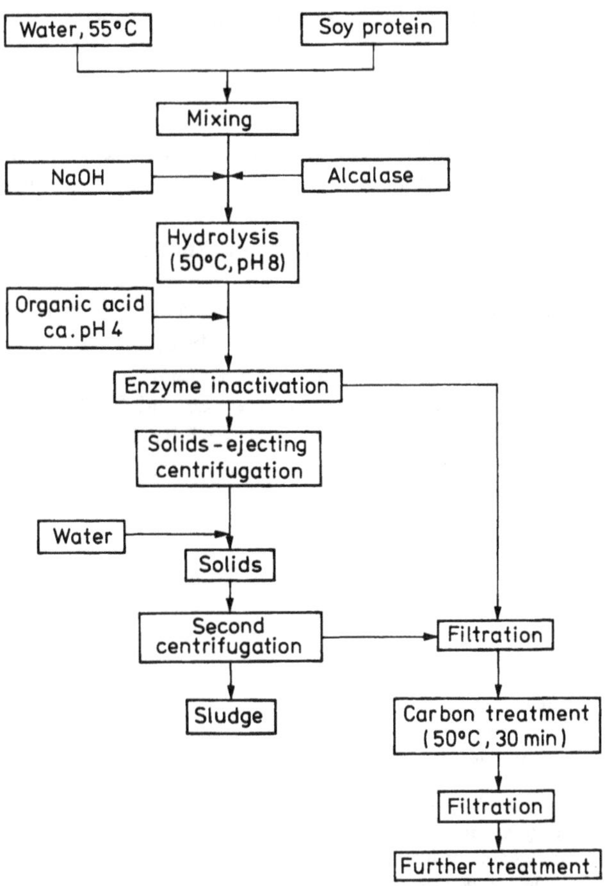

Fig. 7. The enzymatic production of non-bitter soluble protein hydrolysates

6.6 Bakers Dough

6.6.1 The Baking Process

Traditionally, the production of flavour and texture of bread relied on the metabolic conversion and leavening of the bread dough by mixed cultures of yeast and bacteria. The process lasted about 3 h, with the microorganisms producing the alcohol, CO_2, organic acids and other compounds important for the flavour quality of the bread. The whole process required 4.5 h.

Demands for shorter baking times led to the development of a variety of bread-making processes, such as the Chorleywook Bread Process [1,224]. About 75% of the bread sold in the U.K. is made by this process. *Saccharomyces cerevisiae* is the dominant yeast used in baking, and is grown in batch culture on beet or cane molasses. The dough, mixed with yeast, water and salt, is divided into 400 g or 800 g pieces, moulded, then proven and baked. During proving, the yeast metabolises the flour glucose and maltose, producing CO_2 which is trapped in gas cell nuclei arising from the mixing and moulding process. Baking gelatinises the starch, removes water from the dough and inactivates the yeast, resulting in a stable, rigid product. The whole process requires only 1 h. This process also enables the use of mechanical handling systems.

6.6.2 Enzymatic Modification

Wheat is the major source of flour for baking processes. The major constituent of flour is starch, although the unique and variable properties of bakery doughs are due largely to the insoluble gluten proteins of the flour. The time required for dough mixing and loaf characteristics such as volume, symmetry, texture and compressibility of the baked bread depend upon the "strength" of the gluten.

The production of doughs with optimal handling properties and a strict control of mixing times is required due to new methods of bread production, such as mass production, automatic equipment, tight time schedules. Although wheat flour contains a number of naturally occurring enzymes which are essential for the germination and growth of the cereal, their levels are too low to have a significant effect during bread making. As a result, amylases and proteinases are added to the wheat and flour to help in bread production (Table 7).

Table 7. The effects of supplemented amylases and proteinases on the properties of bread

Amylase	Proteinase
Increased bread volume	Improved dough elasticity and texture
Improved crust colour	Increased bread volume
Improved crumb structure	Energy saving
Anti-staling effects	Improved machinability in biscuit production

The amylases convert the starch damaged in the milling process to dextrins and low molecular weight sugars, which are then made available to the yeast during the baking process. Cereals normally contain α- and β-amylases. Adequate levels of β-amylase are present in germinated and ungerminated grains. In contrast, ungerminated grain contains very small amounts of α-amylase, whereas large quantities of the enzyme are present in the germinated grain. Consequently, the levels of α-amylase in flour depend on the conditions of growth and harvesting. If low levels of the enzyme are present the resulting bread will be of inferior quality due to low dextrin production and consequently poor gas production. Hence, low amounts of the enzyme are supplemented to suitable levels with a fungal α-amylase [1)].

Fungal proteinases are added to modify wheat gluten and any milk proteins present (Table 7). The limited proteolysis of doughs improves the elasticity of the gluten which permits easier machining, and consequently results in increased loaf volume, greater symmetry and better grain and texture. The addition of proteinases also reduces the mixing time by about 30% without detrimental effects on the handling properties of the dough [247)]. Correct amounts of proteinase must be added, otherwise slack, porous and sticky dough results if too much enzyme is added, making it difficult to handle [1)]. In the USA about the two-thirds of the white bread made is treated with proteolytic enzymes of *A. oryzae* [248)]. The *Aspergillus* proteinase contains both exo- and endopeptidase activity and has good activity at the acidic pH of the dough. It is also rapidly inactivated during the baking process. The enzymes are supplied in the form of powders mixed with flour [247)]. The neutral proteinase from *B. amyloliquefaciens* may also be used in preparations free from α-amylase [182)].

6.6.3 Other Applications

Fungal and bacterial proteinases are also used in cracker, biscuit and cookie manufacture, improving the extensibility and strength of the dough, allowing it to be rolled very thinly without tearing. The dough also has to be soft to prevent bending and wrinkling of the biscuit in the oven. Very soft and plastic dough is also needed in the precise impression of letters and decoration of biscuits. Bacterial neutral proteinases are most often used for this purpose [182)]. The enzymes are highly specific endopeptidases and are ideal for use on high protein flours.

6.7 Leather Tanning

6.7.1 Skin Composition

The skin, or hide, is composed of three distinct layers: —
1) The epidermis, 2) the corium, or leather skin, and 3) the connective tissues of the under surface.
The hairs are embedded in the epidermis within hair follicles. They are composed of keratin and are fastened in the follicle by proteinaceous material containing large amounts of lysine and arginine.

The corium is composed mainly of bundles of collagen fibres. The bundles are of a regular triple helix structure with intra- and intermolecular bonds, and are stable to proteolysis by the known proteinases. They are interspersed by the reticular tissue containing the protein, elastin, and glycoproteins. The corium layer, which represents the major portion of the skin substance, is firmly bound to the inner tissues of the animal by the connective tissue layer.

Unwanted interfibrillar material of hides is removed in a three step process involving soaking, dehairing and bating.

6.7.2 Soaking

Hides which are to be processed to leather are at first cured to prevent microbial spoilage. Traditional methods such as drying the raw skin in the sun, adding salt to the flesh side and drying, and steeping in a brine bath before drying, are frequently employed.

The cured hides are rehydrated and washed in the tannery to remove dirt and excess fat. The purpose of the soaking is to swell the hide. Traditionally, swelling was achieved by the addition of alkali. Currently, water uptake is increased by the addition of microbial alkaline proteinases to the soak liquor bath. Enzymes used in washing detergents, such as Alcalase 1.5T (Novo) and Milezyme®8X (Miles Laboratories), are often employed as they are compatible with the surfactants and sodium chlorite, which are added to prevent microbial spoilage of the leather. The proteinases are added at the lowest levels required to give the water uptake in the time available [249]. The effect of increasing enzyme concentration on water uptake is illustrated in Fig. 8. Float volumes of 500–1000% are used for stagnant soaks. Using paddle or drum operations, the same effect can be induced with 25-fold lower enzyme concentrations or contact time.

Water uptake and subsequent processing can also be improved by solubilising and washing out fats and gums. This is achieved with pancreatic trypsins [1], which also contain small amounts of amylase and lipase activities. These enzymes may be valuable in the preparation of particularly fatty flesh side skins.

6.7.3 Dehairing and Dewooling

The oldest method used for dehairing is the "sweating process". Raw hide is placed in a steam chamber for several days, enabling the natural bacterial flora to grow and release enzymes which break down the tissues holding the hair in the follicle.

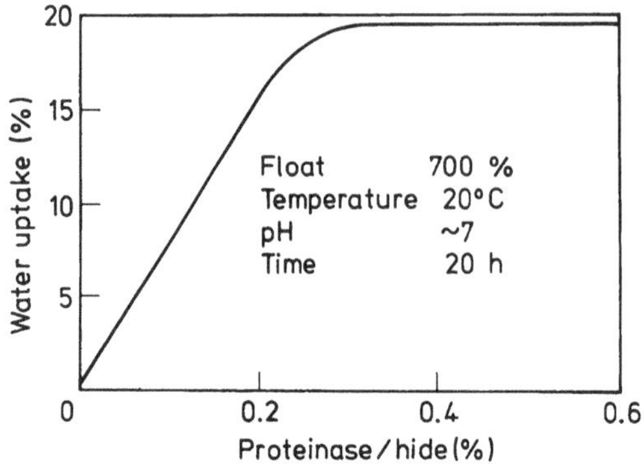

Fig. 8. The effect of bacterial alkaline proteinase concentration on the water uptake during soaking of sheep skins

The major disadvantages of this process are the long time required and the possibility of damage from the bacteria to the collagen fibres and grain.

Dehairing has been improved by the development of extremely alkaline conditions in the soaked and swollen epidermis and corium of the skin. Under these conditions the hair roots and follicles are swollen, which eases the removal of the hairs and soluble debris. The hide is then treated with sulphides to break down bonds of the protein fibrils within the hair and to solubilise the proteins in the hair root. This method is relatively simple and cheap, but causes a build up of hydrogen sulphide in effluents and a waste treatment problem. This has resulted in an increased use of proteinases in hide dehairing, especially with the introduction of highly alkaline proteinases from the *Bacillus* species. In the presence of the proteinases, up to 50% of the lime and sulphide required for a non-enzymatic treatment can be omitted. The process is usually carried out at pH 8–9 for 6 h at 35–40 °C. Milezyme®8X and Novo Unhairing Enzyme® No. 1 are added at up to 0,1%[1,250].

Enzymatic treatment destroys undesirable pigments and increases skin area, thereby shortening or even avoiding the bating stage, providing more uniform dyeing and drastically reducing water pollution.

Coarse-wooled skins are dewooled by painting the flesh side with a dewooling paint (Table 8) and incubating overnight at 20–35 °C. Fine-wooled skins are usually sprinkled evenly over the flesh side with a powdered preparation containing lower levels of proteinases (Table 8) and are then incubated at 25–30 °C for 24 h.

6.7.4 Bating

The collagen of the dehaired hides is delimed and deswollen. The bating process also serves to partially degrade the protein fibres, making them soft, elastic, and enabling them to accept an even dye and demonstrate an acceptable grain. The early methods were based on bacterial proteinases from dog- and bird-faeces. These

Table 8. Composition of the typical dewooling paints used for dewooling a) coarse-wooled and b) fine-wooled skins

a) Coarse-wooled skins	
Hydrated lime	400 g
Sodium chlorite	2.5 g
Proteinase	25 g
Water	1 l
(Used to paint up to 7 skins)	
b) Fine-wooled skins	
Sodium sulphite	27 g
Sodium sulphate	7 g
Ammonium chloride	7 g
Ammonium sulphate	5 g
Proteinase (neutral or alkaline)	4 g

were subsequently replaced with pancreatic trypsin, patented in 1908 by Otto Röhm. Today, as well as trypsin, the proteinases of *A. oryzae*, *B. licheniformis* and *B. amyloliquefaciens* are used [182]. The proteinases are usually supplied as combined preparations [1].

The proteinases used for bating are selected for their specificity toward the various proteins of the hide, especially the matrix proteins, elastin and keratin. The amount of enzyme used depends on whether soft or hard leather is to be produced. For soft and elastic leather strong bates are used; weak bates are used for hard and rigid leather. Accordingly, high proteolytic concentrations are applied for strong bates, and low amounts for weak bates.

Bating conditions, such as temperature, pH, time, chemistry and enzyme concentration, vary according to the type of hide used. Thus, the bating process is a definite craft skill, with the necessary conditions being chosen by an experienced tanner. In general, bating is performed between 25 and 35 °C, and at pH 7.0–9.5.

6.7.5 Latest Developments

Dehairing has been further improved with the use of proteinases specific for the type of skin, since certain enzymes are more effective for some animal species than others. Greater evenness of dehairing may be achieved by a cold soaking to allow penetration of the skin at low enzyme activity, followed by warming to and maintaining at the required temperature for a few hours allowing for even action throughout the skin area.

Dehairing at acid pH using acidic proteinases should offer the advantage of totally eliminating lime and sulphide, thereby further improving the effluent problem.

The use of immobilised dehairing proteinases, with the enzyme bound to a clay which allows it to be mixed with other chemicals needed for dehairing, has been suggested [1].

The increased usage of enzymatic dehairing and dewooling processes has led to an attempt to improve bating performance of the enzymatically-treated skins. The enzymatic processes not only obviate the pollution problems, but also increase energy saving. It is hoped to produce enzymes which in the future can operate at 12–18 °C.

6.8 Protein Hydrolysis

6.8.1 Meat Tenderisation

Proteinases play an important role in meat tenderisation, especially of beef. The toughness of beef varies tremendously depending on the part of carcass from which it came and the condition of the animal [251]. Tenderisation of meat is achieved by the action of endogenous proteinases, especially the lysosomal cathepsins and a neutral metalloproteinase found in or near the myofibrillar proteins [251, 252]. The process requires up to 4 weeks at room temperature, and involves the action of proteinases on the Z-line of the muscular sarcosoma [253]. Endogenous proteinases also play a significant role in meat tenderisation during cooking [254]. Exogenous proteinases, such as papain, ficin, bromelain or fungal proteinases, which act on the connective tissue proteins, elastin and collagen, and on myofibrillar proteins can also be used [253]. The ideal enzymes must be active at pH 4–5 and at low temperatures, and be inactivated at about 50 °C.

Surface tenderisation of meat is usually achieved by sprinkling with a powdered enzyme preparation or by immersion in a proteinase solution. Prepackaged, frozen meats, are dipped into enzyme solutions prior to marketing and can be heated directly from the frozen state. Most of the tenderisation takes place during the cooking process. For surface tenderisation 2% papain or 5% microbial proteinase, especially from *B. subtilis* and *A. oryzae*, is used.

Many tenderisation processes involve the injection of proteinase solutions into the vascular systems of animals 10–30 min before slaughter. The proteolytic enzymes are injected in their inactive oxidised form and are activated by reduction in the freshly slaughtered carcass. The enzymes become distributed throughout the tissues, allowing for a higher percentage of tender meat. Microbial proteinases are rarely used for this purpose since they exhibit too low activity against the connective tissue and cannot be inactivated by oxidation and reactivated by reduction. Active proteinases cannot be injected directly into the vascular system as they will activate the complementary system in the serum via the alternative pathway mechanism. This results in severe shock, haemorrhage and death.

6.8.2 Meat Solubilisation

Soluble meat hydrolysates and meaty flavoured hydrolysates may be produced from bones and offal (raw lung) [255] and from bone residues after mechanical deboning [256] by solubilisation with proteinases. A wide range of proteinases are used to hydrolyse the meat, including many of microbial origin. Previous work has used high temperature optima for meat solubilisation [257,258]. Reaction times of 24 h can be reduced to only 3 h by using lower temperatures. The most appropriate enzyme in terms of cost, solubilisation and other relevant factors appears to be Alcalase [259,260]. In an optimised process 94% solubilisation may be obtained with the Alcalase at pH 8.5 and 55–60 °C [259]. The resulting gravy-like meat slurry is pasteurised for 15 s at 98 °C to inactivate the enzyme, and can be incorporated into canned meat products and soups.

6.8.3 Fish Protein

Proteinases are also used to solubilise fish protein concentrates, which are of high nutritional value, but are insoluble in water as a result of the solvent extraction methods used [261]. The fish protein concentrate is also bland-tasting with poor functionality for most applications.

Fish protein products with interesting organoleptic and functional properties may be obtained by enzyme treatment. The myofibrillar fraction of fish meat, which represents 70–80% of the meat protein, is prepared by washing the meat with 0.1 M NaCl to remove most of the sarcoplasmic proteins [262], then limited proteolysed with a neutral proteinase. The hydrolysate is then precipitated at pH 3 by complexing with sodium hexametaphosphate, washed with isopropanol, neutralised and spray-dried to yield a product of improved emulsifying properties and good water dispersability [263].

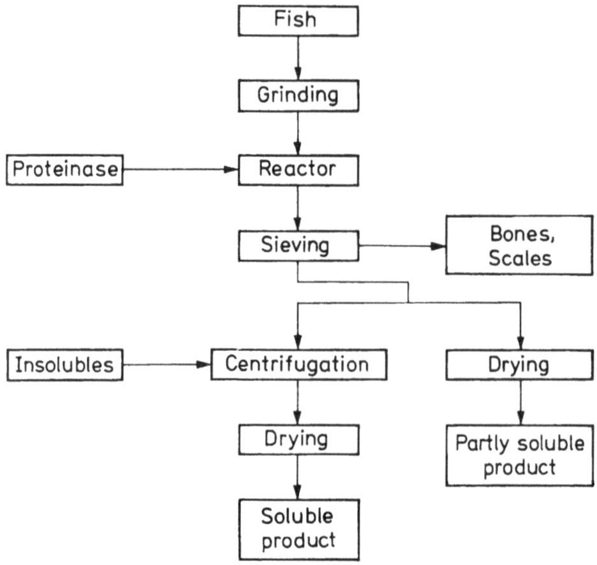

Fig. 9. Enzymatic production of fish protein hydrolysates

Alkaline microbial proteinases, including Alcalase and Neutrase, are also good for fish protein hydrolysis [264]. The process is carried out at 55–60 °C and alkaline pH for 30 min as illustrated in Fig. 9, yielding milk-like products with good nutritional properties [265]. The added proteinases bind to the fish protein before hydrolysing the peptide bonds [266]. Soluble hydrolysates may also be produced by hydrolysing fish fillet waste with Alcalase to 8% degree of hydrolysis. The hydrolysates can be incorporated into fish fillets without decreasing their organoleptic quality [267].

6.8.4 Blood Decolourisation

Blood is an important, but under-utilised, source of food protein [268]. The installation of high capacity systems for the hygienic collection of blood in slaughterhouses has resulted in the availability of food grade quality blood proteins. The blood is separated by centrifugation into plasma and red cell fractions. The blood plasma has excellent emulsifying and heat coagulating properties, making it a useful functional protein in meat emulsions [268]. The red cell fraction, which contains about 75% of the protein in blood, has found little use because of its intense colour. The colour can be removed by acid acetone fractionation, but the handling and recovery of the acetone is an expensive and risky process [269]. Now various enzyme hydrolysis methods are employed [267].

One method involves treatment of alkali-denatured haemoglobin with Alcalase in a membrane reactor, resulting in a light-coloured hydrolysate [270]. Good hydrolysis and decolourisation may also be obtained with fungal proteinases [271], but Alcalase is preferred as it gives the most rapid and thorough hydrolysis of the red cells [267]. Another method is the batch hydrolysis process [272,273], outlined in Fig. 10, which yields a virtually colourless, low molecular weight hydrolysate. The hydrolysate

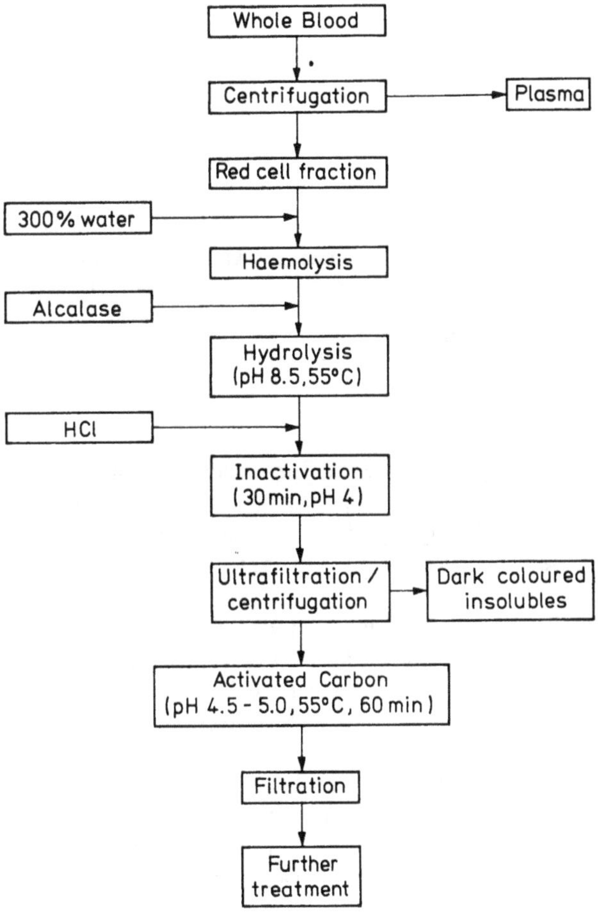

Fig. 10. Enzymatic decolourisation of whole blood

may be remixed with the plasma fraction or be processed and used alone. The hydrolysate can be concentrated to 50% solids content in a falling film evaporator, then spray-dried and agglomerated. The blood cell hydrolysate is soluble and slightly bitter, but can be incorporated at concentrations of 1–3% into meat products, such as sausages and meat pies, without the bitterness being detected.

6.9 Peptide Synthesis

6.9.1 Introduction

The ability of proteinases to catalyse peptide bond formation was demonstrated 50 years ago [274]. However, it is only in the last ten years that enzymatic peptide synthesis has been used as a preparative method for the modification of proteins [275]. Until recently the most commonly used methods for peptide synthesis have been a Reversal of Hydrolysis

$$R_1\text{-COOH} + R_2\text{-NH}_2 \rightleftharpoons R_1\text{-CONH-}R_2 + H_2O$$

b Aminolysis of Esters

$$R_1\text{-OEt} + R_2\text{-NH}_2 \rightarrow R_1\text{-CONH-}R_2 + \text{EtOH}$$

c Aminolysis of Amides

$$R_1\text{-CONH-}R_3 + R_2\text{-NH}_2 \rightarrow R_1\text{-CONH-}R_2 + R_3\text{-NH}_2$$

Fig. 11 a–c. Major routes of proteinase-mediated peptide synthesis. R_1 = peptidyl or protecting group; R_2, R_3 = amino acid residue

chemical [276], using liquid- and solid-phase systems. These methods, though well developed and widely used, have a number of limitations, such as problems of synthesis of long-chain polypeptides, coupling of amino acids or peptides to large protected peptides, and subsequent purification of the finished polypeptide from a mixture of side products [277].

The use of proteinases, usually regarded as degradative enzymes, to effect net synthesis of polypeptides from amino acids or peptides offers several advantages over the chemical methods. The enzymatic reaction can be carried out in aqueous phase, occurs without racemisation, and does not require the protection of side chain functional groups since the reactions are catalysed at the α-amino- and α-carboxyl groups of the amino acids and peptide derivatives. Consequently, a reduced number of side reactions in a synthesis occur, which simplifies the purification procedures and enables the reaction components to be reused. Thus, a total or partial enzymatic synthesis reaction should yield a clean product at a lower cost than the chemical synthesis methods.

Three major routes of proteinase-mediated peptide synthesis are known (Fig. 11), namely;
a) reversal of peptide bond hydrolysis, b) aminolysis of esters, and c) aminolysis of amides (transpeptidation).

6.9.2 Reversal of Hydrolysis

The equilibrium of the reversal of hydrolysis reactions lies to the side of educts. Hence, significant product yields can only be achieved if the equilibrium of the reaction is altered to favour product formation. This can be achieved by;

a) removal of the product from the reaction mixture by precipitation [278],
b) using a biphasic organic/aqueous mixture where the product is more soluble in the organic solvent [279], and
c) addition of high concentrations of cosolvents, such as glycerol or 1,4-butanediol [280].

Peptide bond formation by reversal of hydrolysis can involve stepwise synthesis or fragment condensation mechanisms. The latter is the preferred method for reversal of hydrolysis.

a Noncovalent Complex-forming Fragments

$A + B \rightleftharpoons A{:}B \rightleftharpoons A{-}B$

b Noncomplex-forming Fragments

$A + B \rightleftharpoons A{-}B$

Fig. 12a and b. Types of enzyme fragment condensation reactions. A + B = fragments to be condensed; A:B = noncovalent complex between A and B; A—B = condensation product

Stepwise enzymatic synthesis has been successfully employed in the formation of peptides such as aspartame [281] and angiotensin [282]. Thermolysin was used in the former case; papain, subtilisin BNP′ and a microbial metalloproteinase in the latter. Reversal of hydrolysis has also been used in combination with aminolysis for the stepwise synthesis of Leu- and Met-enkephalin [283] and caerulein [284].

Enzymatic fragment condensation represents the major mechanism of peptide synthesis by reversal of hydrolysis. Two types of condensation have been reported (Fig. 12). These involve the formation of a defined noncovalent complex before resynthesis to the condensation product, or of a product without a defined complex. In the former case, because of the stability of the complex, relatively high condensation yields can be obtained at significantly lower fragment concentrations than in the latter. Enzymatic condensations of noncovalent fragment complexes, with glycerol as cosolvent, have been used for the synthesis of proteins such as bovine pancreatic ribonuclease S [285], staphylococcal nuclease T [286], horse heart cytochrome c [287], and human somatotropin [288].

Despite limitations of the noncomplex-forming condensations, as a consequence of the high fragment concentrations required, the system has been successfully used in the enzymatic formation of human insulin from porcine insulin [289,290]. These limitations can be circumvented by cross-linking the two fragment substrates, thereby converting a condensation from an inter- to an intramolecular one [291]. After condensation the cross-linking group can be removed by reaction with cyanogen bromide. Alternatively, a molecular trap may be used to enhance condensation of the noncomplex-forming fragments. The molecular trap binds specifically to the condensation product and removes it from the chemical equilibrium. A successful application of this method has been the addition of ribonuclease-S-(21–124) protein to a reaction mixture containing two bovine pancreatic ribonuclease S-peptide fragments [292]. The clostripain-catalysed condensation of the two fragments is enhanced by the ribonuclease, which acts as a trap in binding the condensation product but not the fragments, and shifts the equilibrium to favour product formation.

6.9.3 Aminolysis of Amides and Esters

Enzyme-catalysed peptide bond formation by aminolysis of amides and esters involves the transfer of the acyl group of the substrate from an acyl-enzyme complex either to the α-amino group of an amino acid residue or to water. The enzyme reacts rapidly with the substrate to form the acyl-enzyme complex which, in competition with water, reacts with an amino acid-derived nucleophile to form a new peptide bond [275,277]. Partition of the acyl-enzyme intermediate between hydrolysis and aminolysis depends on the concentration of the uncharged form of the amine.

Extensive peptide bond formation is possible provided the product is not hydrolysed by the enzyme to a significant extent. Since at pH 8.5 or above the proteolytic activity of most proteinases is low while their esterase activity is high, strong alkaline conditions may be chosen to favour the formation of relatively stable peptide products [277]. An excess of the amino component and a suitable solvent system also reduce the proteolytic activity [275].

Despite the similarity of the reactions involving aminolysis of amides and esters, the former method has been used little in preparative peptide synthesis. One example of a successful application of aminolysis of amides is the conversion of porcine insulin to human insulin [293,294].

Aminolysis of esters, on the other hand, is the most suited method of peptide bond formation [277,295], since the reaction is fast and is independent of product solubility. Only the esterolytic serine proteinases, such as Subtilisin BNP′, *Streptomyces griseus* chymotrypsin-like proteinase, *Achromobacter* proteinase and carboxypeptidase Y, and cysteine proteinases, such as streptococcal proteinase, can be used with esters as substrates [275,296].

The acyl components used in aminolysis reactions are N-acyl amino acid and peptide alkyl esters; the amine components are the free acids, amides, phenylhydrazides, or alkyl esters of amino acids or peptides [296]. Stepwise synthesis reactions, using amino acid- or peptide esters as amine components, have been used in the synthesis of Met-enkephalin [297], mouse epidermal growth factor [295], and in the modification of the properties of food proteins [245,298–300].

The covalent incorporation of essential amino acids into proteins originally involved the plastein reaction [244,301]. The products, called plasteins and characterised by their low solubility in water, are generally bland in flavour. The process has been used to improve the nutritional properties of food protein [244], to alter the water solubility of soy protein [302,303] and to prepare proteins for special diets [304]. The conventional process requires two enzymatic processes; protein hydrolysis and resynthesis [244]. A one-step process, involving the covalent incorporation into proteins of an L-amino acid ester, is now used to enhance methionine levels in soy protein and flour [245], in improving the surface-active properties of proteins [298,305], for the preparation of antifreeze emulsions [299] and in the production of low-phenylalanine peptides for phenylketonuria patients [300].

6.10 Enzyme Technology

6.10.1 Immobilisation

The field of enzyme immobilisation is a relatively recent biochemical development which has opened many possibilities for enzyme utilisation in industrial processes, stabilising the enzymes against different kinds of inactivation [306–310].

Immobilisation is the conversion of enzymes from a water-soluble, mobile state, to a water-insoluble, immobile state. Over 100 immobilisation techniques have been developed [308]. These can be divided into four groups (Fig. 13);

a) covalent attachment or adsorption of enzymes on solid supports,

b) entrapment of enzymes in polymeric gels,

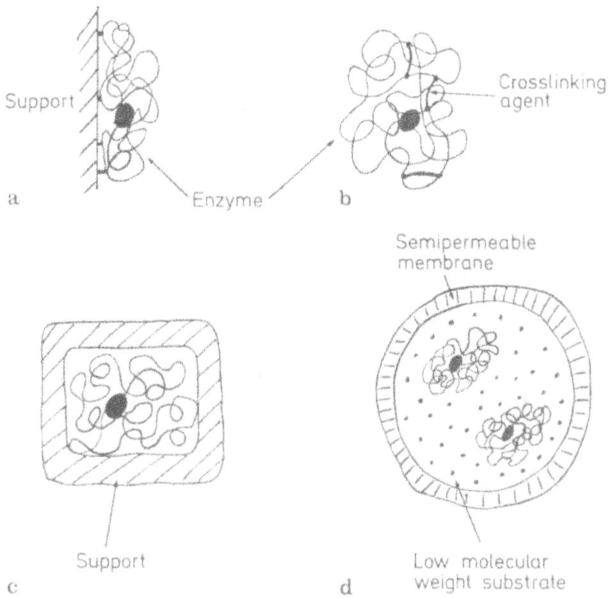

Fig. 13a–d. Methods of enzyme immobilisation. **a** attachment to solid supports; **b** entrapment in gels; **c** cross-linking by bifunctional agents; and **d** microencapsulation

c) cross-linking of enzymes with bifunctional agents, and
d) microencapsulation of enzymes.

Methods involving covalent attachment or covalent cross-linking result in strong chemical bonds between the enzyme and the support. Covalent attachment, however, is an expensive method which may result in significant inactivation of the enzymes due to binding through their active centres. This may be prevented by immobilising in the presence of substrates or cofactors. The adsorption and entrapment methods are simple and efficient, but the enzymes can leak from the supports as the bonds between the enzyme and matrix are weak. Treatment of the adsorbed or entrapped enzymes with cross-linking agents such as glutaraldehyde overcomes this problem [308]. The effects of immobilisation on the properties of the enzyme have been discussed in detail [306,308]. In general, immobilisation of enzymes, such as the proteinases, helps to stabilise them against thermal, operational and storage inactivation brought about by self-digestion, or autolysis, of the free proteinases. Immobilised enzymes can also simplify batch reaction, since the insoluble biocatalysts can be readily removed from the product, simplifying termination of the reaction and subsequent separation of the enzyme. Recovery of the immobilised enzymes enables them to be reused in biocatalytic reactions. Immobilised bacterial and fungal cells can also be used in place of intracellular enzymes which are difficult and expensive to isolate and purify.

Immobilised proteinases have been used in the removal of proteins from antibiotic preparations [311] and in peptide synthesis [312]. In peptide synthesis, a water-wet Amberlite XAD-7 immobilised thermolysin has been employed. Peptide synthesis can also be catalysed by immobilised carboxypeptidase Y and trypsin [313], and *Achromobacter lyticus* proteinase 1 [314]. Proteinase immobilisation by microencapsulation also enhances cheese ripening [225].

6.10.2 Protein Engineering

Despite the success of proteinases and other enzymes in industrial processes, only a handful of the many naturally occurring enzymes are used commercially. This is because many industrial applications are totally different from the physiological role played by the enzymes. In many industrial processes the desired substrate or product is different from the physiological one, and conditions for the reaction, such as pH and temperature, are usually non-physiological. Under these conditions many enzymes are either unstable or function sub-optimally. Improvement of several properties (Table 9) of native enzymes by mutation and selection programmes and chemical modification has been partially effective.

Table 9. Properties of enzymes to be changed for industrial processes

Enzyme turnover (V_{max}) & Michaelis constant (K_m)
Temperature stability and optimum
pH optimum
Stability and activity in organic solvents
Substrate and reaction specificity
Proteinase resistance
Cofactor requirements
Molecular weight and subunit structure
Allosteric regulation

Recent advances in recombinant DNA techniques, site-directed mutagenesis and chemical synthesis of DNA fragments have enabled the systematic alteration in protein structure. As a result, improvements in both the functional and structural characteristics of enzymes are now possible. Protein engineering enables the examination of a known protein structure, the determination of the amino acids to be altered, and the introduction of the necessary changes in the corresponding gene. The modified gene can then be introduced into an appropriate host-vector system, expressed and the protein tested for the desired and predicted characteristics [315]. The ability to engineer a protein is dependent on the knowledge of the three-dimensional structure of the protein [315, 316]. The only method available for determining large protein structure is X-ray crystallography, although two-dimensional NMR techniques can be applied to peptides and proteins up to 10 kDa [317]. Knowledge of the properties and structure of the protein enables prediction of potentially possible and useful modifications.

Protein engineering is still at the stage of learning the structure-function relations in proteins. Subtilisin has been chosen as the model system, since the crystal structures of the enzyme and enzyme-substrate complex are known, the enzyme has been cloned and expressed in large quantities, and the enzyme is of commercial importance. A number of subtle, often predictable alterations in various properties of the enzyme have already been made. Properties such as thermal stability [318], pH profile, catalytic activity and substrate specificity [319–322], as well as reaction type [323], have been altered by modification of charged groups in the active site cleft and on the surface of the subtilisin. The properties of other enzymes, such as yeast triosephosphate isomerase [324] and the tyrosyl-tRNA synthetase of *B. stearothermophilus* [325], have also been studied and redesigned by protein engineering. One interesting and potentially useful modification is the replacement of methionine by phenylalanine at residue 222, next to the active site serine 221. The mutant subtilisin, instead of the normally preferential hydrolytic action, exhibits a preference for transesterification reactions [323]. This characteristic may be useful in the future development of enzymatic protein synthesis (Sect. 6.9). The lessons learned from these studies should enable the creation of novel proteins with improved properties for commercial applications. In fact, enzymes are already being designed for specific commercial uses [323].

6.11 Other Applications

Other commercial applications of microbial proteinases include the modification of products such as beer, wine, cereals, chocolate-cocoa, eggs and animal feeds; the recovery of silver from spent photographic film; and the desizing of fabrics [1, 253, 326]. In brewing, proteinases are involved in body, flavour and nutrient development of the beer, and also aid in filtration and clarification. The enzymes play an important part in the chill proofing of beers, preventing precipitation on cooling at 4 °C. The production of extracellular acid proteinases with properties desirable for chill proofing has recently been reported in two yeasts, *Saccharomycopsis fibuligera* and *Torulopsis magnoliae* [327]. The successful cloning and expression in *Saccharomyces cerevisiae* of the *S. fibuligera* gene coding for the extracellular acid proteinase [328] would appear to open new possibilities in beer manufacture.

Proteinases are also used to modify proteins of cereals to increase the drying rate and improve product handling properties. Hydrolysis of eggs and egg products improves their drying properties, and waste products are converted to animal feeds following proteolytic action.

Microbial proteinases have also found use in dental care and medical therapy and in the pharmaceutical industry. In dental care a mutanase-proteinase mixture has been used to reduce dental plaque and improve the clinical condition of palatal mucosa, or denture stomatitis [329]. The enzymes have been used as digestive aids in gastro-intestinal disorders such as dyspepsia; in the hydrolysis of fibrin and fibrinogen; in the treatment of dermal ulcers and burns; and as potential bacteriocidal agents [13].

Proteinases also play an important role in the removal of protein contaminants, which may cause allergenic reactions, from antibiotic preparations. Although the enzymes are themselves antigenic, they are applied in only small amounts and usually destroy themselves by their own proteolytic action [330]. The unwanted proteins can also be removed by treatment with immobilised proteinases [311]. The immunogeneicity of commercial preparation of benzylpenicillin, ampicillin and related antibiotics can be greatly reduced by treating the compounds with an immobilised proteinase mixture of *Streptomyces griseus*.

7 Conclusion

Proteinases are a very important group of enzymes, both physiologically and economically. In the cell, proteinases catalyse not only the general breakdown of proteins, which along with their synthesis, is important for the regulatory control of the concentrations of proteins. The enzymes also play a significant role in the removal of abnormal proteins; and in numerous limited proteolytic reactions, such as the control of digestion, hormone maturation, viral assembly, immune response, and many other processes, whereby the protein molecules become specifically and selectively modified.

The great diversity of proteolytic action on protein molecules has made the enzymes extremely important in food systems and biotechnology. Proteinases play an important role in the manufacture of detergents and in the processing of foods such as cheese, bread, soy sauce, meat. For economic and technical reasons, microorganisms are the best source of proteinase in many industrial processes and improved technology and enzyme purification should make the microbial enzymes even more attractive.

Recent advances in enzyme immobilisation and protein engineering should help increase the commercial applications of proteinases in processes where the enzymes were previously unstable due to adverse conditions of pH, temperature and presence of oxidising or chelating agents. Immobilised proteinases have already been used in food manufacture, protein synthesis and antibiotic preparation, and their use should be greater in the future. Similarly, recent advances in protein engineering offer the possibility to develop more robust enzymes with the desired properties for industrial processes. Protein engineering should also enable the creation of novel proteins not found in nature.

Enzyme immobilisation and protein engineering techniques are likely to play a significant role in enzymatic protein synthesis reactions. The improvement of current methods in enzymatic peptide synthesis and the isolation and production of proteinases of novel specificities should enable the production of peptides of the highest quality.

The discovery of a large number of highly specific proteinases and the increasing knowledge of their roles in cell metabolism may lead to the introduction of new microbial proteinases in commerce. However, because the costs of using new enzymes for industrial processes are extremely high, introduction of new proteinases is likely to be a slow event, and presently used enzymes are unlikely to be replaced. It seems likely that the use of microbial proteinases in food systems and biotechnology will be increased mainly be developing and modifying existing systems.

In contrast, proteinases are likely to play a major role in advances in genetic engineering [331], where yeast cells are increasingly used for the over-production of required glycoproteins by insertion of the necessary genetic information into yeast. Purification of the product is in many cases hampered by unspecific proteolysis, which can be eliminated by using proteinase-deficient mutants. Alternatively, implantation of genetic information for a non-microbial enzyme, such as calf rennet [228, 332] or human tissue plasminogen activator [333, 334], into a microorganism (*Escherichia coli* or yeast) and subsequent microbial production of that enzyme offers the possibility of microorganisms producing animal and plant proteinases which are in short supply.

Hence, although microbial proteinases already play an important role in industry, their potential is much greater, and their application in future processes is likely to increase.

8 Acknowledgements

The author is indebted to Dr. D. Moore (University of Manchester, U.K.) for encouraging the publication of the review; to Prof. Dr. H. Holzer (University of Freiburg, F.R.G.), Dr. D. Moore and Dr. D. A. Wood (GCRI, Littlehampton, U.K.) for critical assessment of the manuscript; and to Prof. Dr. D. H. Wolf and Dr. M. Müller (University of Freiburg, F.R.G.) for invaluable comments and help with literature survey. The author is also very grateful to Novo Industri A/S, Denmark, for information provided about various applied aspects for the proteinases and for permission to reproduce material. The information provided by Dr. Fred Widmer, Carlsberg Biotechnology, Ltd., Denmark, Henkel KGaA, F.R.G. and Diamalt GmbH, F.R.G., is greatly appreciated. The author would also like to thank Mrs. Ulrike Kopas for typing part of the manuscript, and to Mr. Wolfgang Fritz for help with the figures.

9 Abbreviations

Azocoll	azo dye-bound collagen (an insoluble powdered cowhide containing an attached red dye)
DFP	diisopropyl fluorophosphate
PMSF	phenylmethylsulphonyl fluoride
pCMB	p-chloromercuribenzoate
TLCK	N-tosyl-L-lysine chloromethyl ketone
TPCK	L-1-tosylamide-2-phenylethyl chloromethyl ketone
EDTA	ethylenediaminetetraacetic acid
cAMP	adenosine-3′,5′-cyclic monophosphate
ATP	adenosine triphosphate
DNA	deoxyribonucleic acid
RNA	ribonucleic acid
tRNA	transfer ribonucleic acid
NADP	nicotinamide adenine dinucleotide phosphate
NMR	nuclear magnetic resonance
TAED	tetraacetyl ethylene diamine
CMC	carboxymethyl cellulose
TPP	tripolyphosphate
LAS	linear alkylbenzene sulphonate
CO_2	carbon dioxide

10 References

1. Godfrey, T., Reichelt, J. (eds.): Industrial Enzymology, New York, Nature Press 1983
2. Lorand, L. (ed.): Methods in Enzymology, vol. 45, New York, Academic Press 1976

3. Lorand, L. (ed.): Methods in Enzymology, vol. 80, New York, Academic Press 1981
4. Perlmann, G. E., Lorand, L. (eds.): Methods in Enzymology, vol. 19, New York, Academic Press 1970
5. Bergmeyer, J., Grassl, M. (eds.): Methods in Enzymatic Analysis, vol. 5, Weinheim, Verlag Chemie 1981[3]
6. Achstetter, T., Emter, O., Ehmann, C., Wolf, D. H.: J. Biol. Chem. *259*, 13334 (1984)
7. Law, B. A.: In: Microorganisms and Nitrogen Sources, (Payne, J. W., ed.), p. 381, London, J. Wiley & Sons 1980
8. Cohen, B. L.: ibid., p. 411, London, J. Wiley & Sons 1980
9. Goldberg, A. L., St. John, A. C.: Annu. Rev. Biochem. *45*, 747 (1976)
10. Wolf, D. H.: Adv. Microb. Physiol. *21*, 267 (1980)
11. Segal, H. L., Doyle, D. J. (eds.): Protein Turnover and Lysosome Function, London, Academic Press 1978
12. Stroud, R. M., Krieger, M., Koeppe, R. E. II., Kossiakoff, A. A., Chambers, J. L.: In: Proteases and Biological Control, vol. 2, (Reich, E., Rifkin, D. B., Shaw, E., eds.), p. 13, US, Cold Spring Harbor Lab. 1975
13. North, M. J.: Microbiol. Rev. *46*, 308 (1982)
14. Mizuno, K., Matsuo, H.: Nature *309*, 558 (1984)
15. Wolf, D. H.: Microbiol. Sci. *3*, 107 (1986)
16. Docherty, K., Steiner, D. F.: Annu. Rev. Physiol. *44*, 625 (1982)
17. Hanecak, R., Semler, B. L., Anderson, C. W., Wimmer, E.: Proc. Natl. Acad. Sci. USA *79*, 3973 (1982)
18. Sung, M. T., Cao, T. M., Lischwe, M. A., Coleman, R. T.: J. Biol. Chem. *258*, 8266 (1983)
19. Tweeten, K. A., Bulk, L. A. Jr., Consigli, R. A.: Microbiol. Rev. *45*, 379 (1981)
20. Hirsch, R. L.: ibid. *46*, 71 (1982)
21. Reid, K. B. M., Porter, R. R.: Annu. Rev. Biochem. *50*, 433 (1981)
22. Taylor, P. W.: Microbiol. Rev. *47*, 46 (1983)
23. Holzer, H., Tschesche, H. (eds.): Biological Functions of Proteinases, Berlin, Springer-Verlag 1979
24. Bach, M. K.: Annu. Rev. Microbiol. *36*, 371 (1982)
25. Gordon, A. H., Koj, A. (eds.): The Acute-Phase Response to Injury and Infection, Amsterdam, Elsevier 1985
26. Farach, H. A. Jr., Mundy, D. I., Strittmatter, W. J., Lennarz, W. J.: J. Biol. Chem. *262*, 5483 (1987)
27. Jackson, C. M., Nemerson, Y.: Annu. Rev. Biochem. *49*, 765 (1980)
28. Doolittle, R. F.: Annu. Rev. Biochem. *53*, 195 (1984)
29. Nasjletti, A., Malik, K. U.: Annu. Rev. Physiol. *43*, 597 (1981)
30. Ondetti, M. A., Cushman, D. W.: Annu. Rev. Biochem. *51*, 283 (1982)
31. Margolius, H. S.: Annu. Rev. Physiol. *46*, 309 (1984)
32. Dancer, B. N., Mandelstam, J. J.: J. Bacteriol. *121*, 406 (1975)
33. Wolf, D. H.: Trends Biochem. Sci. *7*, 35 (1982)
34. Postemsky, C. J., Digman, S. S., Setlow, P.: J. Bacteriol. *135*, 841 (1978)
35. Maurizi, M. R., Switzer, R. L.: Curr. Top. Cell. Regul. *16*, 163 (1980)
36. Plaut, A. G.: Annu. Rev. Microbiol. *37*, 603 (1983)
37. Brubaker, R. R.: ibid. *39*, 21 (1985)
38. Roberts, J. W., Roberts, C. W., Craig, N. L.: Proc. Natl. Acad. Sci. USA *75*, 4714 (1978)
39. Gottesman, S.: Cell *23*, 1 (1981)
40. Walker, G. C.: Annu. Rev. Biochem. *54*, 425 (1985)
41. Wong, R. L., Gutowski, J. K., Katz, M., Goldfarb, R. H., Cohen, S.: Proc. Natl. Acad. Sci. USA *84*, 241 (1987)
42. Goldberg, A. L., Dice, J. F.: Annu. Rev. Biochem. *43*, 835 (1974)
43. Rivett, A. J.: Curr. Top. Cell. Regul. *28*, 291 (1986)
44. Segal, H. L., Winkler, J. R.: ibid. *24*, 229 (1984)
45. Goff, S. A., Goldberg, A. L.: Cell *41*, 587 (1985)
46. Wang, S. S., Fried, V. A.: J. Biol. Chem. *262*, 6357 (1987)
47. Switzer, R. L.: Annu. Rev. Microbiol. *31*, 135 (1977)

48. Switzer, R. L., Bond, R. W., Ruppen, M. E., Rosenzweig, S.: Curr. Top. Cell. Regul. *27*, 373 (1985)
49. Kirschke, H., Langner, J., Wiederanders, B., Ansorge, S., Bohley, P., Hanson, H.: In: Intracellular Protein Catabolism II (Turk, V., Marks, N., eds.), p. 299, New York, Plenum Press 1977
50. Voellmy, R., Murakami, K., Goldberg, A. L.: In: Limited Proteolysis in Microorganisms (Cohen, G. N., Holzer, H., eds.), p. 7, Washington, US Department of Health, Education and Welfare 1979
51. Hershko, A., Ciechanover, A.: Annu. Rev. Biochem. *51*, 335 (1982)
52. Larimore, F. S., Waxman, L., Goldberg, A. L.: J. Biol. Chem. *257*, 4187 (1982)
53. Chung, C. H., Goldberg, A. L.: Proc. Natl. Acad. Sci. USA *78*, 4931 (1981)
54. Chung, C. H., Waxman, L., Goldberg, A. L.: J. Biol. Chem. *258*, 215 (1983)
55. Goff, S. A., Goldberg, A. L.: ibid. *262*, 4508 (1987)
56. Maurizi, M. R., Trisler, P., Gottesman, S.: J. Bacteriol. *164*, 1124 (1985)
57. Katajama-Fujimura, Y., Gottesman, S., Maurizi, M. R.: J. Biol. Chem. *262*, 4477 (1987)
58. Goldberg, A. L., Waxman, L.: ibid. *260*, 12029 (1985)
59. Menon, A. S., Waxman, L., Goldberg, A. L.: ibid. *262*, 722 (1987)
60. Ciehanover, A., Finley, D., Varshavsky, A.: Cell *37*, 57 (1984)
61. Finley, D., Varshavsky, A.: Trends Biochem. Sci. *10*, 343 (1985)
62. Hershko, A., Leshinsky, E., Ganoth, D., Heller, H.: Proc. Natl. Acad. Sci. USA *81*, 1619 (1984)
63. Kornberg, A., Spudich, J. A., Nelson, D. L., Deutscher, M. P.: Annu. Rev. Biochem. *37*, 51 (1968)
64. Esposito, R. E., Klapholz, S.: In: The Molecular Biology of the Yeast Saccharomyces. Life Cycle and Inheritance (Strathern, J. N., Jones, E. W., Broach, J. R., eds.), U.S., Cold Spring Harbor Laboratory 1981
65. Wood, D. A.: Bull. Br. Mycol. Soc. *12*, 120 (1978)
66. Wolf, D. H., Holzer, H.: In: Microorganisms and Nitrogen Sources (Payne, J. W., ed.), p. 431, London, J. Wiley & Sons 1980
67. Wright, B. E., Anderson, M. L.: Biochim. Biophys. Acta *43*, 62 (1960)
68. Schwalb, M. N.: J. Biol. Chem. *252*, 8435 (1977)
69. Burnett, T. J., Shankweiler, G. W., Hageman, J. H.: J. Bacteriol. *165*, 139 (1986)
70. Murao, S., Kameda, M., Nishino, T.: Agric. Biol. Chem. *43*, 1997 (1979)
71. Shimizu, Y., Nishino, T., Murao, S.: ibid. *47*, 1775 (1983)
72. Shimizu, Y., Nishino, T., Murao, S.: ibid. *48*, 3109 (1984)
73. Li, E., Yousten, A. A.: Appl. Microbiol. *30*, 354 (1975)
74. Murao, S., Shimizu, Y., Kameda, M., Nishino, T.: Agric. Biol. Chem. *46*, 3075 (1982)
75. Kawamura, F., Doi, R. H.: J. Bacteriol. *160*, 442 (1984)
76. Yang, M. Y., Ferrari, E., Henner, D. J.: ibid. *160*, 15 (1984)
77. Yanagita, T., Nomachi, Y.: J. Gen. Appl. Microbiol. *13*, 227 (1967)
78. Setlow, P.: In: Sporulation and Germination (Levinson, H. S., Sonenshein, A. L., Tipper, D. J., eds.), p. 13, Washington D.C., American Society for Microbiology 1981
79. Setlow, P., Primus, G.: J. Biol. Chem. *250*, 623 (1975)
80. Setlow, P., Ozols, J.: ibid. *255*, 8413 (1980)
81. Connors, M. J., Setlow, P.: J. Bacteriol. *161*, 333 (1985)
82. Leighton, T. J., Stock, J. J.: ibid. *101*, 931 (1970)
83. Begueret, J., Bernet, J.: Nature (New Biol.) *243*, 94 (1973)
84. Jackson, D. P., Cotter, D. A.: Arch. Microbiol. *137*, 205 (1984)
85. O'Day, D. H.: J. Bacteriol. *125*, 8 (1976)
86. Hershko, A., Fry, M.: Annu. Rev. Biochem. *44*, 775 (1975)
87. Reich, E., Rifkin, D. B., Shaw, E. (eds.): Proteases and Biological Control, USA, Cold Spring Harbor Laboratory 1975
88. Cabib, E.: Trends Biochem. Sci. *1*, 275 (1976)
89. Cabib, E., Roberts, R., Bowers, B.: Annu. Rev. Biochem. *51*, 763 (1982)
90. Cohen, G. N., Holzer, H. (eds.): Limited Proteolysis in Microorganisms, Washington, U.S. Department of Health, Education and Welfare 1979
91. Moreno, S., Galvagno, M. A., Passeron, S.: Arch. Biochem. Biophys. *214*, 573 (1982)
92. Jones, E. W.: Annu. Rev. Genet. *18*, 233 (1984)
93. Kurjan, J., Herskowitz, I.: Cell *30*, 933 (1982)

94. Mizuno, K., Nakamura, T., Takada, K., Sakakibara, S., Matsuo, H.: Biochem. Biophys. Res. Commun. *144*, 807 (1987)
95. Müller, M., Müller, H., Holzer, H.: J. Biol. Chem. *256*, 723 (1981)
96. Funayama, S., Gancedo, J. M., Gancedo, C.: Eur. J. Biochem. *109*, 61 (1980)
97. Witt, I., Kronau, R., Holzer, H.: Biochim. Biophys. Acta *128*, 63 (1966)
98. Holzer, H., Betz, H., Ebner, E.: Curr. Top. Cell. Regul. *9*, 103 (1975)
99. Müller, D., Holzer, H.: Biochem. Biophys. Res. Commun. *103*, 926 (1981)
100. Schäfer, W., Kalisz, H. M., Holzer, H.: Biochim. Biophys. Acta *925*, 150 (1987)
101. Ogasahara, K., Tsunasawa, S., Soda, Y., Yutani, K., Sugino, Y.: Eur. J. Biochem. *150*, 17 (1985)
102. Holzer, H., Katsunuma, T., Schott, E. G., Ferguson, A. R., Hasilik, A., Betz, H.: Adv. Enzym. Regul. *11*, 53 (1973)
103. Betz, H., Weiser, U.: Eur. J. Biochem. *70*, 385 (1976)
104. Waindle, L. M., Switzer, R. L.: J. Bacteriol. *114*, 517 (1973)
105. Ruppen, M. E., Switzer, R. L.: ibid. *155*, 56 (1983)
106. Leitzmann, C., Bernlohr, R. W.: Biochim. Biophys. Acta *151*, 461 (1968)
107. Maurizi, M. R., Brabson, J. S., Switzer, R. L.: J. Biol. Chem. *253*, 5585 (1978)
108. Ruppen, M. E., Switzer, R. L.: ibid. *258*, 2843 (1983)
109. Levine, R. L.: Curr. Top. Cell. Regul. *27*, 305 (1985)
110. Jany, K.-D., Nitsche, E.: Hoppe-Seyler's Z. Physiol. Chem. *364*, 839 (1983)
111. Yu, P.-H., Kula, M.-R., Tsai, H.: Eur. J. Biochem. *32*, 129 (1973)
112. Rouget, P., Chapeville, F.: ibid. *23*, 459 (1971)
113. Sadoff, H. L., Celikkol, E., Engelbrecht, H. L.: Proc. Natl. Acad. Sci. USA *66*, 844 (1970)
114. Orrego, C., Kerjan, P., Manca de Nadra, M. C., Szulmajster, J.: J. Bacteriol. *116*, 636 (1973)
115. Blobel, G., Dobberstein, B.: J. Cell. Biol. *67*, 852 (1975)
116. von Heijne, G.: J. Mol. Biol. *184*, 99 (1985)
117. Michaelis, S., Beckwith, J.: Annu. Rev. Microbiol. *36*, 435 (1982)
118. von Heijne, G.: J. Mol. Biol. *192*, 287 (1986)
119. Meyer, D. I., Krause, E., Dobberstein, B.: Nature *297*, 647 (1982)
120. Walter, P., Gilmore, R., Blobel, G.: Cell *38*, 5 (1984)
121. Lingappa, V. R., Devillers-Thiery, A., Blobel, G.: Proc. Natl. Acad. Sci. USA *74*, 2432 (1977)
122. Pohlner, J., Halter, R., Beyreuther, K., Meyer, T. F.: Nature *325*, 458 (1987)
123. Matsubara, H., Feder, J.: In: The Enzymes, vol. 3, (Boyer, P. D., ed.), p. 721, New York, Academic Press 1971[3]
124. Kellam, S. J., Walker, J. M.: Trans. Biochem. Soc. *15*, 520 (1987)
125. Ferrari, E., Howard, S. M. H., Hoch, J. A.: J. Bacteriol. *166*, 173 (1986)
126. Zlotnik, H., Schramm, U. L., Buckley, H. R.: ibid. *157*, 627 (1984)
127. Shepherd, M. G., Poulter, R. T. M., Sullivan, P. A.: Annu. Rev. Microbiol. *39*, 579 (1985)
128. Honda, T., Booth, B. A., Boesman-Finkelstein, M., Finkelstein, R. A.: Infect. Immun. *55*, 451 (1987)
129. Korant, B. D.: Presented at Labatt's Workshop on Proteinases, Ontario, Canada 1986
130. Dean, D. D., Domnas, A. J.: Arch. Microbiol. *136*, 212 (1983)
131. Frandsen, E. V. G., Reinholdt, J., Kilian, M.: Infect. Immun. *55*, 631 (1987)
132. Mortensen, S. B., Kilian, M.: ibid. 45, 550 (1984)
133. MacDonald, F.: Sabouraudia *22*, 79 (1984)
134. Hislop, E. C., Paver, J. L., Keon, J. P. R.: J. Gen. Microbiol. *128*, 799 (1982)
135. Cazzulo, J. J.: Comp. Biochem. Physiol. *79B*, 309 (1984)
136. Holzer, H., Heinrich, P. C.: Annu. Rev. Biochem. *49*, 63 (1980)
137. Motizuki, M., Mitsui, K., Endo, Y., Tsurugi, K.: Eur. J. Biochem. *158*, 345 (1986)
138. Hay, R., Böhni, P., Gasser, S.: Biochim. Biophys. Acta *779*, 65 (1984)
139. Yasuhara, T., Ohashi, A.: Biochem. Biophys. Res. Commun. *144*, 277 (1987)
140. Tsujita, Y., Endo, A.: Biochem. (Tokyo) *88*, 113 (1980)
141. Regnier, P., Thang, M. N.: FEBS Lett. *102*, 291 (1979)
142. Wiemken, A., Schellenberg, M., Urech, K.: Arch. Microbiol. *123*, 23 (1979)
143. Bertini, F., Brandes, D., Buctow, D. E.: Biochim. Biophys. Acta *107*, 171 (1965)
144. Achstetter, T., Wolf, D. H.: Yeast *1*, 139 (1985)
145. Croall, D. E., De Martino, G. N.: Biochim. Biophys. Acta *788*, 348 (1984)

146. St. John, A. C., Goldberg, A. L.: J. Biol. Chem. *253*, 2705 (1978)
147. Katunuma, N., Umezawa, H., Holzer, H. (eds.): Proteinase Inhibitors: Medical and Biological Aspects, Tokyo, Japan Scientific Societies Press and Springer Verlag 1983
148. Holzer, H.: In: Gene Expression in Yeast (Korhola, M., Vaisanen, E., eds.), p. 147, Proc. of Alko Yeast Symp., Helsinki, Foundation for Biotechnol & Industrial Fermentation Research 1 1983
149. Glenn, A. R.: Annu. Rev. Microbiol. *30*, 41 (1976)
150. Hageman, J. H., Shankweiler, G. W., Wall, P. R., Franich, K., McCowan, G. W., Cauble, S. M., Grajeda, J., Quinones, C.: J. Bacteriol. *160*, 438 (1984)
151. Abbas-Ali, B., Coleman, G.: Trans. Biochem. Soc. *5*, 420 (1977)
152. Cohen, B. L.: J. Gen. Microbiol. *77*, 521 (1973)
153. Cohen, B. L.: Trans. Br. Mycol. Soc. *76*, 447 (1981)
154. Hanson, M. A., Marzluf, G. A.: Proc. Natl. Acad. Sci. USA *72*, 1240 (1975)
155. Cohen, B. L., Drucker, H.: Arch. Biochem. Biophys. *182*, 601 (1977)
156. Abbott, R. J., Marzluf, G. A.: J. Bacteriol. *159*, 505 (1984)
157. Lindberg, R. A., Rhodes, W. G., Eirich, L. D., Drucker, H.: ibid. *150*, 1103 (1982)
158. Grove, G., Marzluf, G. A.: J. Biol. Chem. *256*, 463 (1981)
159. Kalisz, H. M., Wood, D. A., Moore, D.: Trans. Br. Mycol. Soc. *88*, 221 (1987)
160. Kalisz, H. M., Wood, D. A., Moore, D.: In preparation
161. Enzyme Nomenclature: Recommendations of the International Union of Biochemistry, New York, Academic Press 1978
162. Morihara, K.: Adv. Enzymol. *41*, 179 (1974)
163. Kraut, J.: Annu. Rev. Biochem. *46*, 331 (1977)
164. Govind, N. S., Merta, B., Sharma, M., Modi, V. V.: Phytochem. *20*, 2483 (1981)
165. Boguslawski, G., Shultz, J. L., Yehle, C. O.: Anal. Biochem. *132*, 41 (1983)
166. Lindberg, R. A., Eirich, L. D., Price, J. S., Wolfinbarger, L. Jr., Drucker, H.: J. Biol. Chem. *256*, 811 (1981)
167. Eriksson, K.-E., Pettersson, B.: Eur. J. Biochem. *124*, 635 (1982)
168. Fujimura, S., Nakamura, T.: Infect. Immun. *55*, 716 (1987)
169. North, M. J., Whyte, A.: J. Gen. Microbiol. *130*, 123 (1984)
170. North, M. J., Roper, A. M., Walker, M.: FEMS Microbiol. Lett. *21*, 175 (1984)
171. Oda, K., Terashita, T., Kono, M., Murao, S.: Agric. Biol. Chem. *45*, 2339 (1981)
172. Pearl, L. H.: FEBS Lett. *214*, 8 (1987)
173. Rhodes, W. G., Lindberg, R. A., Drucker, H.: Arch. Biochem. Biophys. *223*, 514 (1983)
174. Gripon, J. C., Auberger, B., Lenoir, J.: Internat. J. Biochem. *12*, 451 (1980)
175. Beck, C. I., Scott, D.: Adv. Chem. Series *136*, 1 (1974)
176. Eveleigh, D. E., Montenecourt, B. S.: Adv. Appl. Microbiol. *25*, 57 (1979)
177. Aunstrup, K., Andresen, O., Falch, E. A., Nielsen, T. K.: In: Microbial Technology, vol. 1 (Reppler, H. J., Perlman, D., eds.), p. 281, New York, Academic Press 1979[2]
178. Layman, P. L.: Chem. Eng. News, p. 11, Sept. 15 (1986)
179. Aunstrup, K., Outtrup, H.: British Patent 1,303,633 (1973)
180. Jolliffe, L. K., Doyle, R. J., Streips, U. N.: J. Bacteriol. *141*, 1199 (1980)
181. Anon.: McGraw-Hill's Biotechnol. Newswatch *3*, 1 (1983)
182. Austrup, K.: In: Economic Microbiology, vol. 5 (Rose, A. H., ed.), p. 49, London, Academic Press 1980
183. Nakadai, T., Nasuno, S., Iguchi, N.: Agric. Biol. Chem. *37*, 2703 (1973)
184. Nasuno, S.: J. Gen. Microbiol. *70*, 29 (1972)
185. Kohman, H. A., Irwin, R., Stateler, E. S.: US Patent 1,654,176 (1927)
186. Arima, K., Yu, J., Iwasaka, S., Tamura, G.: Appl. Microbiol. *16*, 1727 (1968)
187. Aunstrup, K.: British Patent 1,108,287 (1968)
188. Sardinas, J. L.: US Patent 3,275,453 (1966)
189. Yoshida, F., Ichishima, E.: US Patent 3,149,051 (1964)
190. Klapper, B. F., Jameson, D. M., Mayer, R. M.: Biochim. Biophys. Acta *304*, 505 (1973)
191. Aunstrup, K., Andresen, O., Outtrup, H.: US Patent 3,723,250 (1973)
192. Horikoshi, K., Ikeda, Y.: US Patent 4,052,262 (1977)
193. Keay, L., Anderson, R. G.: US Patent 3,592,737 (1971)
194. Endo, S.: J. Ferment. Technol. *40*, 346 (1962)

195. Keay, L., Moseley, M. H., Andersen, R. G., O'Connor, R. J., Wildi, B. S.: Biotechnol. Bioeng. Symposium *3*, 63 (1972)
196. Murray, E. D., Prince, M. P.: US Patent 3,507,750 (1970)
197. Novo Industri A/S: personal communication
198. Christensen, P. N., Thomsen, K., Branner, S.: Presented at the 2nd World Conference on Detergents, Montreux, Switzerland, Novo Industri A/S No. A-05953 (1986)
199. Stinson, S. V.: Chem. Eng. News, p. 21, Jan. 26 (1987)
200. Novo Industri A/S: Information Bulletin No. B 247d–GB 2000, Bagsvaerd, Denmark, Novo Industri A/S (1986)
201. Eidgenössische Material Prüfungsanstalt (EMPA), St. Gallen, Switzerland
202. Novo Industri A/S: Information Bulletin No. B 157e–GB 2000, Bagsvaerd, Denmark, Novo Industri A/S (1985)
203. Jensen, G.: Lecture at the XVth Spanish CED/AID Meeting, Novo Industri A/S File No. A-05845 (1984)
204. Novo Industri A/S: Information Bulletin Nos. B338c–GB 2500 and B 346c–GB 1500, Bagsvaerd, Denmark, Novo Industri A/S (1987)
205. Jensen, G.: Presented at the 5th Yugoslav Symposium on Surface Active Substances, Ohrid, Novo Industri A/S File No. A-05753 (1981)
206. Novo Industri A/S: Information Bulletin No. B 398a–GB 300, Bagsvaerd, Denmark, Novo Industri A/S (1987)
207. Weismantel, G.: Chemical Week, p. 9, Jan. 22 (1986)
208. US Dept. of Agriculture: Cheese Varieties and Descriptions, Washington D.C., USDA 1978
209. Sternberg, M.: Adv. Appl. Microbiol. *20*, 135 (1976)
210. Sardinas, J. L.: ibid. *15*, 39 (1972)
211. Galloway, J. H., Crawford, R. J. M.: In: Microbiology of Fermented Foods, vol. 1 (Wood, B. J. B., ed.), p. 111, London, Elsevier Applied Science Publishers 1985
212. Olson, N. F.: In: Microbial Technology — Fermentation Technology, vol. 2 (Peppler, H. J., Perlman, D., eds.), p. 39, New York, Academic Press 1979[2]
213. Green, M. L.: J. Dairy Res. *44*, 159 (1977)
214. Rymaszewski, J., Poznanski, S., Reps, A., Ichilczyk, J.: Milchwissenschaft *28*, 779 (1973)
215. Grappin, R., Rank, T. C., Olson, N. F.: J. Dairy Sci. *68*, 531 (1985)
216. Chapman, H. R., Sharpe, M. E.: In: Dairy Microbiology, vol. 2 (Robinson, R. K., ed.), p. 157, Lonson, Applied Science Publishers 1981
217. Lawrence, R. C., Thomas, T. D., Terzaghi, B. E.: J. Dairy Res. *43*, 141 (1976)
218. Mathesson, A. R.: N. Z. J. Dairy Sci. Technol. *16*, 33 (1981)
219. Yun, S.-E., Ohmiya, K., Shimizu, S.: Agric. Biol. Chem. *46*, 443 (1982)
220. Richardson, B. C., Pearce, K. N.: N. Z. J. Dairy Sci. Technol. *16*, 209 (1981)
221. Trieu-Cuot, P., Gripon, J.-C.: J. Dairy Res. *49*, 501 (1982)
222. Marcos, A., Esteban, M. A., Leon, F., Fernandez-Salguero, J.: J. Dairy Sci. *62*, 892 (1979)
223. Law, B. A., Kolstad, J.: Antonie van Leeuwenhoek J. Microbiol. Serol. *49*, 225 (1983)
224. Beech, G. A., Melvin, M. A., Taggart, J.: In: Biotechnology. Principles and Applications (Higgins, I. J., Best, D. J., Jones, J., eds.), p. 73, Oxford, Blackwell Sci. Publ. 1986
225. Anon.: Chem. & Industry, p. 832, Dec. 15 (1986)
226. Hori, T.: J. Food Sci. *50*, 911 (1985)
227. Marconi, W.: Enzyme Engineering — Future Directions (Wingard, L. B., Jr., Berezin, I. V., Klyosou, A. A., eds.), p. 465, London, Plenum Press 1980
228. McCaman, M. T., Andrews, W. H., Files, J. G.: J. Biotechnol. *2*, 177 (1985)
229. Yokotsuka, T.: In: Microbiology of Fermented Foods, vol. 1, (Wood, B. J. B. ed.), p. 197, London, Elsevier Applied Sci. Publ. 1985
230. Yong, F. M., Wood, B. J. B.: Adv. Appl. Microbiol. *17*, 157 (1974)
231. Hesseltine, C. W.: Annu. Rev. Microbiol. *37*, 575 (1983)
232. Wang, H. L., Hesseltine, C. W.: In: Microbial Technology — Fermentation Technology, vol. 2, (Peppler, H. J., Perlman, D., eds.), p. 95, London, Academic Press 1979[2]
233. Kundu, A. K., Manna, S.: Appl. Microbiol. *30*, 507 (1975)
234. Impoolsup, A., Bhumiratana, A., Flegel, T. W.: Appl. Environ, Microbiol. *42*, 619 (1981)
235. Nakadai, T.: J. Japan Soy Sauce Res. Inst. *3*, 99 (1977)

236. Bhumiratana, A., Flegel, T. W., Glirisukon, T., Somporan, W.: Appl. Environ. Microbiol. *39*, 430 (1980)
237. Mohri, M., Matsushita, S.: J. Agric. Food Chem. *32*, 486 (1984)
238. Fuke, Y., Matsuoka, H.: J. Food Sci. *49*, 312 (1984)
239. Park, Y. W., Kusakabe, I., Kobayashi, H., Murakami, K.: Agric Biol. Chem. *49*, 3215 (1985)
240. Babji, A. S., Adnan, A., Aminah, A.: Pertanika *7*, 1 (1985)
241. Chiba, H., Takahashi, N., Sasaki, R.: Agric. Biol. Chem. *43*, 1883 (1979)
242. Ochiai, K., Kamata, Y., Shibasaki, K.: ibid. *46*, 91 (1982)
243. Fuke, Y., Matsuoka, H.: Nippon Shokuhin Kogyo Gahkaishi *27*, 275 (1980)
244. Fujimaki, M., Arai, S., Yamashita, M.: Adv. Chem. Series *160*, 156 (1977)
245. Yamashita, M., Arai, S., Amano, Y., Fujimaki, M.: Agric. Biol. Chem. *43*, 1065 (1979)
246. Adler-Nissen, J.: Annales Nutrition l'Alimentation *32*, 205 (1978)
247. Reed, G. (ed.): Enzymes in Food Processing, New York, Academic Press 1966
248. Barrett, F. F.: In: Enzymes in Food Processing (Reed, G., ed.), p. 301, New York, Academic Press 1975
249. Novo Industri A/S: Information Bulletin No. B 378–GB, Bagsvaerd, Denmark, Novo Industri A/S (1986)
250. Alexander, K. T. W., Haines, B. M., Walker, M. P.: JALCA *81*, 85 (1986)
251. Dransfield, E., Etherington, D.: In: Enzymes and Food Processing (Birch, G. G., Blakebrough, N., Parker, K. J., eds.), p. 177, London, Applied Sci. Publ. 1981
252. Busch, W. A., Stromer, M. H., Goll, D. E., Suzuki, A.: J. Cell Biol. *52*, 367 (1972)
253. Whitaker, J. R.: Adv. Chem. Series *160*, 95 (1977)
254. Patestos, N. P., Harrington, M. G.: Trans. Biochem. Soc. *15*, 266 (1987)
255. Moss, V. G., Trautman, J. C.: US Patent 3,692,538 (1972)
256. Behnke, U., Ackermann, E., Ruttloff, H.: Nahrung *28*, 397 (1984)
257. Novo Industri A/S: Information Bulletin No. 163–GB, Bagsvaerd, Denmark, Novo Industri A/S (1978)
258. Hale, M. B.: In: Making Fish Protein Concentrates by Enzymatic Hydrolysis. Report NMFS SSRF-657, Seattle, USA: National Marine Fisheries Service 1972
259. O'Meara, G. M., Munro, P. A.: Enzyme Microbiol. Technol. *6*, 181 (1984)
260. Ledward, D. A., Lawrie, R. A.: J. Chem. Technol. Biotechnol. *34B*, 223 (1984)
261. Hevia, P., Whitaker, J. R., Olcott, H. S.: J. Agric. Food Chem. *24*, 383 (1976)
262. Groninger, H. S. Jr.: ibid. *21*, 978 (1973)
263. Spinelli, J., Koury, B. J., Miller, R.: J. Food Sci. *37*, 604 (1972)
264. Lalasidis, G., Bostrom, S., Sjoberg, L.-B.: J. Agric. Food Chem. *26*, 751 (1978)
265. Yancz, E., Bollester, D., Monckeberg, F.: J. Food Sci. *41*, 1289 (1976)
266. Mohr, V.: In: Biochemical Aspects of New Protein Food, vol. 44 (Adler-Nissen, J., Eggum, B. O., Munck, L., Olsen, H. S., eds.), p. 53, Oxford, Pergamon Press 1977
267. Adler-Nissen, J. (ed.): Enzymic Hydrolysis of Food Proteins, London, Elsevier Applied Sci. Publ. 1986
268. Wismer-Pederson, J.: Food Technol., p. 76, Aug. (1979)
269. Tybor, P. T., Dill, C. W., Landmann, W. A.: J. Food Sci. *40*, 155 (1975)
270. Stachowicz, K. J., Eriksson, C. E., Tjelle, S.: Amer. Chem. Soc. Symp. Series *47*, 295 (1977)
271. Drepper, G., Drepper, K., Ludwig-Busch, H.: Fleischwirtschaft *61*, 1393 (1981)
272. Hold-Christensen, V., Adler-Nissen, J., Olsen, H. S.: US Patent 4,262,022 (1981)
273. Novo Industri A/S: Information Bulletin No. B 225c–GB 1000, Bagsvaerd, Denmark, Novo Industri A/S 1984
274. Bergmann, M., Fraenkel-Conrat, H.: J. Biol. Chem. *124*, 321 (1937)
275. Fruton, J. S.: Adv. Enzymol. *53*, 239 (1982)
276. Gross, E., Meienhofer, J. (eds.): The Peptides, vols 1–3, New York, Academic Press 1979–1981
277. Chaiken, I. M., Komoriya, A., Ohno, M., Widmer, F.: Appl. Biochem. Biotechnol. *7*, 385 (1982)
278. Tanimoto, S.-Y., Yamashita, M., Arai, S., Fujimaki, M.: Agric. Biol. Chem. *36*, 1595 (1972)
279. Seminov, A. N., Berezin, I. V., Mashnek, K.: Biotech. Bioeng. *23*, 355 (1981)
280. Homandberg, G. A., Mattis, J. A., Laskowski, M. Jr.: Biochem. *17*, 5220 (1978)
281. Oyama, K., Nishimura, S., Nonaka, Y., Kihara, K., Mashimoto, T.: J. Org. Chem. *46*, 5242 (1981)

282. Isowa, Y., Ohmori, M., Satoh, M., Mori, K.: Bull. Chem. Soc. Japan *50*, 2766 (1977)
283. Kullmann, W.: J. Biol. Chem. *256*, 1301 (1981)
284. Takai, H., Sakato, K., Nakamizo, N., Isowa, Y.: In: Peptide Chemistry 1980 (Okawa, K. ed.), p. 213, Osaka, Protein Research Foundation 1981
285. Homandberg, G. A., Laskowski, M. Jr.: Biochem. *18*, 586 (1979)
286. Komoriya, A., Homandberg, G. A., Chaiken, I. M.: Int. J. Pept. Prot. Res. *16*, 433 (1980)
287. Juillerat, M., Homandberg, G. A.: ibid. *18*, 335 (1981)
288. Graf, L., Li, C. H.: Proc. Natl. Acad. Sci. USA *78*, 6135 (1981)
289. Inouye, K., Watanabe, K., Morihara, K., Tochino, Y., Kanaya, T., Emma, J., Sakakibara, S.: J. Amer. Chem. Soc. *101*, 751 (1979)
290. Tager, H., Thomas, N., Assoian, R., Rubenstein, A., Salkow, M., Olefsky, J., Kaiser, E. T.: Proc. Natl. Acad. Sci. USA *77*, 3181 (1980)
291. Chu, S.-C., Wang, C.-C., Brandenburg, D.: Hoppe-Seyler's Z. Physiol. Chem. *362*, 647 (1981)
292. Homandberg, G. A., Komoriya, A., Chaiken, I. M.: Biochem. *21*, 3385 (1982)
293. Breddam, K., Johansen, J. T.: Carlsberg Res. Commun. *49*. 463 (1984)
294. Markussen, J., Volund, A.: In: Enzymes in Organic Synthesis, Ciba Foundation Symposium 111, p. 188, London, Pitman (1985)
295. Widmer, F., Bayne, S., Houen, G., Moss, B. A., Rigby, R. D., Whittaker, R. G., Johansen, J. T.: In: Peptides 1984 (Ragnarsson, U., ed.), p. 193, Stockholm, Almqvist 1984
296. Widmer, F., Johansen, J. T.: In: Synthetic Peptides in Biology and Medicine (Alitalo, K., Partanen, P., Vaheri, A., eds.), p. 79, Amsterdam, Elsevier Sci. Publ. B. V. 1985
297. Widmer, F., Breddam, K., Johansen, J. T.: In: Peptides 1980 (Brunfeldt, K., ed.), p. 46, Copenhagen, Scriptor 1981
298. Arai, S., Watanabe, M.: Agric. Biol. Chem. *44*, 1979 (1980)
299. Watanabe, M., Tsuji, R. F., Hirao, N., Arai, S.: ibid. *49*, 3291 (1985)
300. Arai, S., Maeda, A., Matsumura, M., Hirao, N., Watanabe, M.: ibid. *50*, 2929 (1986)
301. Whitaker, J. R.: Adv. Chem. Series *198*, 57 (1982)
302. Yamashita, M., Arai, S., Kokubo, S., Aso, K., Fujimaki, M.: J. Agric. Food Chem. *23*, 27 (1975)
303. Aso, H., Kimura, H., Watanabe, M., Arai, S.: Agric. Biol. Chem. *49*, 1649 (1985)
304. Yamashita, M., Arai, S., Fujimaki, M.: J. Food Sci. *41*, 1029 (1976)
305. Shimada, A., Yazawa, E., Arai, S.: Agric. Biol. Chem. *46*, 173 (1982)
306. Klibanov, A. M.: Anal. Biochem. *93*, 1 (1979)
307. Martinek, K., Mozhaev, V. V.: Adv. Enzymol. *57*, 179 (1985)
308. Klibanov, A. M.: Science *219*, 722 (1983)
309. Linko, Y.-Y., Linko, P.: In: Biocatalysts in Organic Syntheses (Tramper, J., van der Plas, H. C., Linko, P., eds.), p. 159, Amsterdam, Elsevier Sci. Publ. 1985
310. Wingard, L. B. Jr., Katchalski-Katzir, E., Goldstein, L. (eds.): Applied Biochemistry and Bioengineering, vol. 1, New York, Academic Press 1976
311. Shaltiel, S., Sela, M.: US Patent 3, 770, 584 (1973)
312. Oyama, K., Nishimura, S., Nonaka, Y., Kihara, K., Mashimoto, T.: J. Org. Chem. *46*, 5241 (1981)
313. Royer, G. P.: Canadian Patent 1,114,762 (1981)
314. Oka, T., Muneyuki, R., Morihara, K.: In: Peptides — Structure and Function, 8th Amer. Peptide Symp. (Hruby, V. J., Rich, D. H., eds.), p. 199, Rockford, Pierce 1983
315. Witholt, B., Lageveen, R. G., Kok, M.: In: Biocatalysts in Organic Syntheses (Tramper, J., van der Plas, H. C., Linko, P., eds.), p. 239, Amsterdam, Elsevier Sci. Publ. 1985
316. Ulmer, K. M.: Science *219*, 666 (1983)
317. Wüthrich, K., Wider, G., Wagner, G., Braun, W.: J. Mol. Biol. *155*, 311 (1982)
318. Pantoliano, M. W., Ladner, R. C., Bryan, P. N., Rollence, M. L., Wood, J. F., Poulos, T. L.: Biochem. *26*, 2077 (1987)
319. Russell, A. J., Thomas, P. G., Fersht, A. R.: J. Mol. Biol. *193*, 803 (1987)
320. Russell, A. J., Fersht, A. R.: Nature *328*, 496 (1987)
321. Baum, R.: Chem. Eng. News, p. 23, Oct. 13 (1986)
322. Estell, D. A., Graycar, T. P., Miller, J. V., Powers, D. B., Burnier, J. P., Ng, P. G., Wells, J. A.: Science *233*, 659 (1986)
323. Van Brunt, J.: Bio/Technol. *4*, 928 (1986)

324. Ahern, T. J., Casal, J. I., Petsko, G. A., Klibanov, A. M.: Proc. Natl. Acad. Sci. USA *84*, 675 (1987)
325. Leatherbarrow, R. J., Wells, T. N. C., Fersht, A. R.: In: Enzymes as Catalysts in Organic Synthesis (Schneider, M. P., ed.), p. 311, Lancaster, UK, D. Reidel Publ. Co. 1986
326. Underkofler, L. A.: In: Industrial Microbiology (Miller, B. M., Litsky, W. eds.), p. 128, New York, McGraw-Hill 1976
327. Bilinski, C. A., Russell, I., Stewart, G. G.: Presented at the EMBO Workshop on Proteinases, Freiburg, FRG 1986
328. Yamashita, I., Hirata, D., Machida, M., Fukui, S.: Agric. Biol. Chem. *50*, 109 (1986)
329. Budtz-Joergensen, E., Kelstrup, J.: Scand. J. Dent. Res. *85*, 209 (1977)
330. Abbott, B. J.: Adv. Appl. Microbiol. *20*, 203 (1976)
331. Brake, A. J., Merryweather, J. P., Coit, D. G., Heberlein, U. A., Masiarz, F. R., Mullenbach, G. T., Urdea, M. S., Valenzuela, P., Barr, P. J.: Proc. Natl. Acad. Sci. USA *81*, 4642 (1984)
332. Hill, J. C., Stout, J. T., Rownd, R. H.: In: Abstracts, 12th International Conference on Yeast Genetics and Molecular Biology, p. 140, Edinburgh 1984
333. Gill, G. S., Thomsen, D. R., Bruce, B. J., Marotti, K. R.: ibid., p. 137
334. Meyhack, B., Barth, F., Grütter, M., Heim, J., Treichler, H. J., Hinnen, A.: ibid., p. 138

The Amidases from a *Brevibacterium* Strain: Study and Applications

M. Maestracci, K. Bui, A. Thiéry, A. Arnaud* and P. Galzy
Chaire de Génétique et Microbiologie Ecole Nationale Supérieure Agronomique de Montpellier Place Viala 34060 Montpellier Cedex/France

* To whom correspondence should be addressed.

In the first part of this paper different microbial enzyme systems able to hydrolyze the true amide function are reviewed, then some enzyme systems are described in detail: aminopeptidases, amidases of *Aspergillus nidulans* and of *Pseudomonas aeruginosa*.

Following consideration of analytical techniques (NMR, GLC, TLC, colorimetric assay of the ammonium ion) useful in the study of amidases, the second part of the paper deals with the amidases of *Brevibacterium* sp. R312 and their applications. Selection conditions of the strain, demonstrations of the different amidase systems, and complete studies of the L-α-aminoamidase and the wide spectrum amidase are described. Some examples of applications and future developments are also given.

1 Introduction

Many commercially interesting organic acids are prepared from nitriles by either alkali or acid chemical hydrolysis [1]. These acids could also be produced by biological hydrolysis of nitriles. Such a hydrolysis involves in most cases two enzyme systems: a nitrile hydratase and an amidase. The direct transformation: nitrile → acid is quite rare [2-7]. The study of amidases involves:
— the improvement of the bioconversion of nitriles into acids,
— the stereospecificity of the molecule, and in particular the possibility of producing optically active α-aminoacids,
— the hydrolysis process of amides under mild conditions.

In the present paper, we first review the enzyme systems able to transform amides and the techniques used in the study of biological hydrolysis of these compounds. The second part of the paper deals specifically with the amidase systems of *Brevibacterium* and their applications.

2 Microbial Enzymes Hydrolyzing Amides

Only enzymes able to hydrolyze the true amide function are considered. They must be able to catalyze the following reaction:

$$R{-}C\genfrac{}{}{0pt}{}{\nearrow O}{\searrow NH_2} \xrightarrow{+H_2O} R{-}COOH + NH_3$$

2.1 The Aminopeptidases (L-Leucyl-peptide Hydrolase E.C. 3.4.1.1 and Amino-acyl-oligopeptide Hydrolase E.C. 3.4.1.2)

These enzymes hydrolyze some peptides, but they are also able to hydrolyze the amide function of some L-α-aminoamides [8,9]. The study of this type of enzyme will be presented in Sect. 3.

2.2 Asparaginase (L-Asparagine Amidohydrolase E.C. 3.5.1.1)

Catalyzed reaction:

$$\text{L-asparagine} \xrightarrow{+H_2O} \text{L-aspartate} + NH_3$$

This enzyme is found in *Escherichia coli* [10] and *Pseudomonas* sp. [11] in particular. The enzyme from *Pseudomonas* can also hydrolyze L-glutamine. Two asparaginases have been described for *E. coli* (E.C. 1; E.C. 2). Whilst E.C. 2 hydrolyzes both L-asparagine and L-glutamine, E.C. 1 hydrolyzes only L-asparagine.

2.3 Glutaminase (L-Glutamine Amidohydrolase E.C. 3.5.1.2)

This enzyme is found in *E. coli* [12] and effects the following reaction:

$$\text{L-glutamine} \xrightarrow{+H_2O} \text{L-glutamate} + NH_3$$

It also hydrolyzes the α-methyl-DL-glutamine, but does not attack L-isoglutamine, D and L homoglutamine, α-methyl-DL-asparagine, L-isoasparagine, L-leucinamide, DL-proline amide, DL-alanine amide, L-phenylalanine amide, L-tyrosine amide and the aliphatic amides.

2.4 ω-Amidase (ω-Amidodicarboxylate Amidohydrolase, E.C. 3.5.1.3)

Catalyzed reaction:

$$\omega\text{-amido dicarboxylic acid} \xrightarrow{+H_2O} \text{dicarboxylate} + NH_3$$

This enzyme is encountered in some bacteria: *Bacillus subtilis* and *Thermus aquaticus* [13]. This type of enzyme hydrolyzes a fairly large number of ω-amido dicarboxylic acids. The 5-methyl-2-oxoglutarate is the substrate most rapidly hydrolyzed.

2.5 Amidase (Acylamide Amidohydrolase E.C. 3.5.1.4)

Catalyzed reaction:

$$\text{monocarboxylic acid amide} \xrightarrow{+H_2O} \text{monocarboxylate} + NH_3$$

This type of enzyme will be discussed in detail in Sect. 4.

2.6 Urease (Urea Amidohydrolase E.C. 3.5.1.5)

Catalyzed reaction:

$$\text{urea} \xrightarrow{+H_2O} CO_2 + 2\ NH_3$$

This enzyme was obtained from *Bacillus pasteurii* [14] and from *Corynebacterium renale* [15] in particular. It is specific for urea.

2.7 Biotinidase (Biotin-amide Amidohydrolase E.C. 3.5.1.12)

Catalyzed reaction:

$$\text{biotin amide} \xrightarrow{+H_2O} \text{biotin} + NH_3$$

This enzyme can be found in *Streptococcus faecalis* [16]. It is able to hydrolyze esters and amides of biotin.

2.8 Nicotinamide Deaminase (Nicotinamide Amidohydrolase E.C. 3.5.1.a)

Catalyzed reaction:

$$\text{nicotinamide} \xrightarrow{+H_2O} \text{nicotinate} + NH_3$$

This enzyme is found in *Torula cremoris* [17] and *Mycobacterium avium* [18–21]. This type of enzyme is very specific and only hydrolyzes nicotinamide.

2.9 5-Aminovaleramidase (5-Aminovaleramide Amidohydrolase, E.C. 3.5.1.a.a)

Catalyzed reaction:

$$\text{5-aminovaleramide} \xrightarrow{+H_2O} \text{5-aminovalerate} + NH_3$$

This enzyme is found in *Pseudomonas putida* [22], and can attack 4-aminobutyramide and 6-aminocaproamide. However, it does not hydrolyze butyramide or valeramide.

2.10 Allantoinase (Allantoin Amidohydrolase E.C. 3.5.25)

Catalyzed reaction:

$$\text{allantoin} \xrightarrow{+H_2O} \text{allantoate}$$

This enzyme is found in *Streptococcus allantoicus*, *Arthrobacter allantoicus*, *E. coli*, *Pseudomonas acidovorans*, and *Pseudomonas fluorescens* [23]. Besides allantoin, this type of enzyme can also hydrolyze methylallantoin, and sometimes 5-aminohydantoin. However, 3-methylallantoin and 5-acetylallantoin are not degraded by this enzyme.

3 Microbial Aminopeptidases

Table 1 summarizes the properties of the main aminopeptidases described.

Most of these aminopeptidases have an alkali optimum pH, usually above 8.0. These enzymes can be found in different locations in the cell. It is noteworthy that they are rarely found in ribosomes. The optimal temperature of aminopeptidases is high (usually over 50 °C). Some are stimulated by higher temperatures and their stability during heating is good. These enzymes are usually metalloenzymes and are inhibited by heavy metals but activated by Mg^{2+} and Mn^{2+}, Zn^{2+}, or Co^{2+}. The functions at the active site most often described are the —SH groups, the S—S bridges and the —OH groups of serine.

Table 1. Properties of the main microbial aminopeptidases described

Authors	Organism	Optimum pH	Localization	Optimal temperature and thermal denaturation	Action of Cations	Study of the active site
Berger et al. [24]	12 bacterial species	8 to 9	Soluble			
Brown [25]	*E. coli* B	7.8–9			Activation by Mn^{2+}	Presence of —SH
Chapuis and Zuber [26]	*Talaromyces duponti*	6.9		Activation by high temperatures Stable at temperatures >60 °C	Activation by Co^{2+}	Inhibition by EDTA
De Marco and Dick [27]	Several microorganisms		Ribosomal	Stable up to 75 °C	Activation by Mg^{2+} and Zn^{2+}	
Foissy [28, 29]	*Brevibacterium linens*	9.6	Exocellular	Optimum 27 °C Stable up to 50 °C	Activation by Co^{2+}	
Hayman et al. [30]	*E. coli* B		Soluble		Activation by Zn^{2+}; Mn^{2+} and Co^{2+}	Inhibition by EDTA
Ivanova et al. [31]	*Aspergillus oryzae*	9	Soluble		Inhibition by Hg^{2+}; Pb^{2+}; Cu^{2+}; Zn^{2+}; Fe^{2+}; Mn^{2+}	Presence of S—S bridges Absence of —SH and serine —OH
Koelsch and Hanson [32]	*Micrococcus lysodeikticus*	9.2–10	Soluble		Activation by Mn^{2+}; Mg^{2+} and Fe^{2+}	Absence of —OH and —SH Inhibition by EDTA

Lehmann and Uhlig [33]	*Aspergillus*	8			Activation by Co^{2+}	Inhibition by EDTA
Masuda et al. [34]	Yeast hydrolysate	8	Soluble	Optimum 40 °C	Activation by Zn^{2+}; Co^{2+}; Mn^{2+}	Inhibition by EDTA Presence of —SH
Matheson [35] Matheson et al. [36] Matheson and Tsai [37] Tsai and Matheson [38]	*E. coli* B	7.5–9.2	Ribosomal		Activation by Mg^{2+}	Resistance to trypsin
Mimamiura et al. [39,40]	*Bacillus subtilis*	7.5–8.5	Soluble	Stable up to 60 °C	Activation by Mn^{2+}; Co^{2+} and Mg^{2+}	Inhibition by EDTA Absence of —SH
Morihara and Tsuzuki [41]	*Streptomyces sioyaensis*	8–9	Exocellular		Activation by Co^{2+}	
Nakadai et al. [42]	*Aspergillus oryzae*	Depends on substrate	Exocellular	Optimum 60 °C Resists up to 70 °C	No activation by cations	Inhibition EDTA Presence of S—S bridges Absence of —SH Inhibition by DFP (presence of —OH)
Nakadai et al. [43]	*Aspergillus oryzae*	8	Exocellular	Optimum 50 °C Resists up to 70 °C Activated by temperature	Activation by Co^{2+}	Inhibition by EDTA Presence of S—S bridges Absence of —OH

Table 1 (continued)

Authors	Organism	Optimum pH	Localization	Optimal temperature and thermal denaturation	Action of Cations	Study of the active site
Nakadai et al.[44]	*Aspergillus oryzae*		Exocellular	Optimum 50 °C Resists up to 70 °C	Activation by Co^{2+}; Mn^{2+}; Zn^{2+}; Ca^{2+}	Inhibition by EDTA Presence of S—S Absence of —OH
Nakadai et al.[45]	*Aspergillus oryzae*	7	Exocellular	Optimum 40 °C		No effect by EDTA Presence of —OH Absence of S—S and —SH
Prescott and Wilkes[46]	*Aeromonas*	8–8.5	Exocellular	Resists up to 70 °C	Activation by Zn^{2+} and Mn^{2+}	Inhibition by EDTA
Roncari and Zuber[47] Roncari et al.[48]	*Bacillus stearothermophilus*	9.2–9.4 depending on substrate	Exocellular	Optimum 90 °C Activation by temperature Denaturation after 15 h at 80 °C	Activation by Co^{2+}	Inhibition by EDTA
Ruffin et al.[49]	*Keratomyces ajelloi*	9.35	Exocellular			Inhibition by EDTA Presence of S—S Absence of —SH

Sugiura et al. [50–52]	*Aspergillus japonica*	8	Exocellular	Optimum 50 °C Denatured after 15 min at 50 °C	Activation by Zn^{++}; Co^{++}	Inhibition by EDTA No effect by proteases Presence of S—S Absence of —SH; —OH
Uwajima et al. [53]	*Streptomyces peptidofaciens*		Exocellular	Denatured after 60 min at 70 °C	Activation by Ca^{++}	Inhibition by EDTA
Vogt [54]	*E. coli*	9.5	Soluble	Stable up to 75 °C	Activation by Mg^{++} and Mn^{++} Inactivation by Zn^{++}	Inactivation by EDTA
Wagner et al. [55]	*Bacillus subtilis*	8	Exocellular		Activation by Co^{++}	

4 The Acylamide Amidohydrolases

The acylamide amidohydrolases seem to be widespread among protista organisms: bacteria [56], yeasts [57] and molds [58-60]. As far as bacteria are concerned, amidases can be found in most species: *Corynebacterium* [61,62], *Mycobacterium* [63-66], *Pseudomonas* [56,67-71], *Bacillus* [72,73], *Micrococcus* [72], *Brevibacterium* [72], *Nocardia* [74], *Streptomyces* [75], *Arthrobacter* [76], *Rhodococcus* [77] and *Alcaligenes* [78].

The main properties of some of these amidases are reported in Table 2. It is noteworthy that these enzymes are all inducible and that their specificity varies from one species to another.

In the next two sections we will consider in detail the results of Hynes and his co-

Table 2. Properties of some amidases described in the literature

Strain	Amides, substrates of the amidase	Amides, not substrates of the amidase	Regulation mode	Amide most rapidly hydrolyzed	References
Alcaligenes entrophus	Acetamide Propionamide Butyramide Valeramide	Formamide	Inducible Catabolic repression by fructose and succinate	Valeramide	Friedrich and Mitrenga [78]
Arthrobacter sp.	Acetamide Acrylamide Propionamide	Formamide Butyramide Valeramide Isobutyramide Methacrylamide Malonamide Succinamide Lactamide Glycinamide Benzamide Urea Nicotinamide	Inducible	Acrylamide	Asano et al. [76]
Bacillus sp.	Acetamide Propionamide Fluoro-acetamide Formamide Glycinamide Butyramide Benzamide Nicotinamide Valeramide Cyanoacetamide Thioacetamide Urea Thiourea	Dimethyl-formamide Dimethyl-acetamide *N*-methyl-acetamide	Inducible Catabolic repression by glucose	Acetamide	Thalenfeld and Grossowicz [73]

Table 2 (continued)

Strain	Amides, substrates of the amidase	Amides, not substrates of the amidase	Regulation mode	Amide most rapidly hydrolyzed	References
Mycobacterium smegmatis	Formamide Acetamide Propionamide Butyramide Valeramide Acrylamide Isobutyramide Glycolamide Lactamide β-hydroxy-propionamide Benzamide Nicotinamide	*N*-methyl-acetamide *N*-acetyl-acetamide	Inducible	Formamide	Draper [64]
Rhodococcus sp.	Formamide Acetamide Propionamide Butyramide Acrylamide Nicotinamide	Benzamide Phenyl-acetamide Malonamide	Inducible Catabolic repression by succinate Repression by ammonium ion	Propionamide	Miller and Gray [77]

workers on the amidases from *Aspergillus nidulans* and those of Clarke and her co-workers on the amidases of *Pseudomonas*.

5 Amidases of *Aspergillus nidulans*

Four different enzymes have been shown to exist in *A. nidulans*.

5.1 A Formamidase

This formamidase is specific for formamide and glycinamide (Table 3). It is not inducible. Its biosynthesis is repressed by the ammonium ion, glutamate and glutamine, no matter which substrate is used as the carbon source. This formamidase is slightly affected by catabolic repression by glucose [58–60].

5.2 An Acetamidase

This acetamidase is able to hydrolyze many aliphatic amides [59] (Table 3). The biosynthesis of this enzyme is induced by:

Table 3. Activity spectrum of amidases from *Aspergillus nidulans*

Amides	Formamidase	Acetamidase	Wide spectrum amidase
Formamide	+	—	—
Acetamide	—	+	—
Propionamide	—	+	—
Butyramide	—	+	+
Valeramide	—	+	+
Hexamide	—	+	+
Hydroxyacetamide		+	—
Glycinamide	+	—	—
Benzamide	—	—	+
Phenylacetamide	—	—	+
Nicotinamide	nd	—	—
β-hydroxypropionamide	—	—	—
Lactamide	—	—	—
Fumaramide	—	+	—
Malonamide	—	—	—

+ : substrate of the enzyme
— : not substrate of the enzyme
nd: not determined

— sources of acetyl-CoA: acetamide, acetate, ethanol, L-threonin [79],
— benzamide and benzoate [58,60],
— β-alanine and β-amino acids [80],
— β-hydroxypropionamide [58].

The effects of these compounds are practically cumulative. Understanding of the induction mechanism was improved by the selection of regulation mutants which showed that the acetamidase is submitted to a complex regulation process involving several independent systems. For example, damage to the genes Amd R and Gat A specifically affected the induction by ω-amino acids [80]. As for the Amd 19 mutant, the induction by acetyl-CoA sources was specifically modified [81-83].

Unlike the above formamidase, the biosynthesis of the acetamidase from *A. nidulans* is strongly subject to catabolic repression by glucose [59]. The biosynthesis of this enzyme is also repressed by the ammonium ion and again in contrast to the formamidase, this repression depends on the source of nitrogen in the medium. In the presence of glucose, repression by the ammonium ion occurs, whereas, when the carbon source is acetamide or acetate, the ammonium ion only weakly represses the biosynthesis of acetamidase [59]. Due to this regulation, the metabolism of carbon does not interfere with the use of acetamide as a nitrogen source. Inversely, repression by the ammonium ion does not prevent acetamide being used as a carbon source.

In order to localize and map the structural Amd S gene coding for the acetamidase, Hynes [83] isolated a large number of mutants defective for this enzyme. Most of these mutants were obtained following mutagenesis with diepoxyoctane. Hynes [83] found up to 14 possible mutation sites within the Amd S gene.

The mutation Amd I 18 resulted in a two- or three-fold increase in the synthesis of acetamidase under any culture conditions. The mutation Amd I 9 increased the

Table 4. Different genes involved in the control of the biosynthesis of the acetamidase in *A. nidulans* [81]

Genes	Function	References
amdS	Structural gene of acetamidase	Hynes and Pateman [84] Dunsmuir and Hynes [85]
amd19	Control gene for induction by the sources of acetyl-CoA. Near the gene AmdS	Hynes [79-86]
facA, facB, facC	Control of enzymes involved in the assimilation of acetate	Apirion [87], Armitt et al. [88] Hynes [79]
amdR (intA)	Control of induction of acetamidase by ω-amino acids	Hynes and Pateman [84], Arst [89], Hynes [81]
gatA	Structural gene of ω-amino-acid transaminase. Lesion of this gene results in the accumulation of the inducer ω-amino acid	Arst [89], Hynes [81]
amdA	Unknown	Arst and Cove [90], Hynes [81]
areA	Gene responsible for the repression of acetamidase by the ammonium ion	Hynes [91], Hynes [92]

induction effect of acetamide, acetate and acetyl-CoA sources. These two mutations, which involved the control area of the synthesis of this enzyme, are situated close to the structural gene Amd S [82,83]. These results demonstrated the separation of the structural zone and the control zone of a gene in an eukaryotic organism. Hynes [81] isolated different mutants and identified 9 genes involved in the control of the structure of the amidase (Table 4).

Table 5. Activity spectra and regulation modes of the three main amidases from *Aspergillus nidulans*

Enzymes	Structural gene	Main substrates	Regulation modes
Formamidase	fmdS	Formamide	Non-inducible Repression by the ammonium ion Weak catabolic repression by glucose
Acetamidase	amdS	Acetamide Propionamide Butyramide Valeramide Hexamide	Inducible Repression by the ammonium ion Catabolic repression by glucose
Amidase with a wide spectrum	gmdS	Benzamide Phenylacetamide Butyramide Valeramide Hexamide	Non-inducible Repression by the ammonium ion

5.3 An Amidase Described by Hynes as Having a Wide Spectrum

This wide spectrum amidase could hydrolyze benzamide, phenylacetamide and some aromatic compounds (Table 3). This enzyme is not inducible and its biosynthesis is repressed by the ammonium ion, glutamate and glutamine [60].

5.4 An Amidase Specific for Valeramide and Hexamide

The existence of this enzyme was demonstrated by Hynes [60], however no further work was reported.

Table 5 summarizes the reports on the main three amidases of *A. nidulans*.

6 Amidases of *Pseudomonas*

6.1 The Different Types of Amidases in *Pseudomonas sp.*

Clarke and her co-workers studied the amidase systems of several species of *Pseudomonas* sp. [56,69,93,94] and concluded that there are two types of amidases in the wild type strains:

a) The aliphatic amidases which hydrolyze acetamide. These enzymes are present in *P. aeruginosa* strains and in some *P. putida*, *P. cepacia* and *P. acidovorans* strains.

b) The phenylacetamidases specific for phenylacetamide. This type of enzyme is only found in *P. putida*, *P. cepacia* and *P. acidovorans*.

Using immunodiffusion tests, the amidases of *Pseudomonas aeruginosa* were shown to be identical to each other. On the other hand, *P. putida* and *P. acidovorans* species contain an amidase which only showed a partial homology with that of the *P. aeruginosa* strains. The *P. cepacia* produced an amidase quite different from those described above.

6.2 The Acylamide Amidohydrolase from *Pseudomonas aeruginosa* 8602/A

6.2.1 Specificity and Regulation Mode

This enzyme has a restricted activity spectrum and only hydrolyzes amides with four carbon atoms or less. However, it can be induced by a large number of amides or amide analogues [95] (Table 6).

Similarly to the amidase of *Aspergillus nidulans*, the biosynthesis of the amidase of *Pseudomonas aeruginosa* is subject to catabolic repression [96]. Cyanoacetamide strongly represses the biosynthesis of this enzyme. Using ^{14}C- marked *N*-acetylacetamide as inducer, cyanoacetamide was shown to act on the control system for the biosynthesis of the amidase and not on the penetration into the cell by the inducer.

Other work showed that the hydrolysis of aliphatic amides involved the following enzyme and genetic systems:

a) A constitutive permease which concentrates amides within the cell. Acetamide was concentrated 100-fold and acetylacetamide 80-fold [97].

Table 6. Amides and amide analogues as substrates and inducers of the amidase from *Pseudomonas aeruginosa*

Amide	Substrate	Inducer
Formamide	+	—
Acetamide	+	+
Propionamide	+	+
Butyramide	+	—
Isobutyramide	—	—
Valeramide	—	nd
Hexamide	—	nd
N-Methylformamide	—	+
N-Ethylformamide	—	+
N-Methylacetamide	—	+
N-Ethylacetamide	—	+
N-Acetylacetamide	—	+
N-Phenylacetamide	—	—
N,N'-Dimethylacetamide	—	—
N-Methylpropionamide	—	+
N-Ethylpropionamide	—	+
Cyanoacetamide	—	—
Iodoacetamide	—	nd
Glycine amide	—	—
Hydroxyacetamide	+	+
Acrylamide	+	nd
Lactamide	—	+
Urea	—	—
Thioacetamide	—	—

nd: not determined

b) A regulation gene controlling the biosynthesis of the amidase [98,99]. The presence of this gene accounts for the induction and repression results obtained.
c) A structural gene for the biosynthesis of the amidase [100].

Using phages for transduction tests, the regulation gene and the structural gene for the amidase were shown to be very close together despite their distinctive properties. A high frequency of simultaneous transduction was obtained for these two genes [101].

6.2.2 Mutants of *Pseudomonas aeruginosa* 8602/A

Clarke and her co-workers isolated a large number of regulation mutants and those with modified amidases.
1) Regulation mutants.
a) Constitutive mutants [101].
b) Magno-constitutive mutants. Without induction these strains could produce as much amidase as the wild type with induction, but they could not produce more.
c) Semi-constitutive mutants. These strains produce without induction 10%–50% of the amount of amidase produced by the induced wild type. When induced, these strains produce the same amount of enzyme as the wild type. It is noteworthy that the amidase activity is nil or very weak in the wild type when not induced.
d) Mutants with an amidase inducible by formamide [101].

e) Mutants with the biosynthesis of amidase not repressed by butyramide [56,102,103]. Butyramide is a very poor substrate, a non-inducer and an inhibitor for the biosynthesis of amidase in the wild type. The mutants described were isolated from strains with a constitutive amidase and were shown to have an amidase no longer repressed by butyramide.
f) Mutants insensitive to catabolic repression [101].
2) Mutants with modified amidases.
a) Amidase defective mutants [104]. Fluoroacetamide was used for the selection of these strains. Following mutagenesis, the cells were spread plated on a minimum medium containing pyruvate and fluoroacetamide. Cells with amidase activity hydrolyze fluoroacetamide into toxic fluoroacetic acid forming very small colonies. Cells which are amidase defective could grow on this medium and form visible colonies.
b) Mutants able to utilize butyramide as nitrogen source [102]. These strains were isolated from amidase constitutive strains and could hydrolyze butyramide more rapidly than their parent strains.
c) Mutants able to utilize valeramide as a nitrogen source [102]. These strains were also isolated from amidase constitutive parents and could hydrolyze valeramide. However, some could no longer hydrolyze acetamide.
d) Mutants able to utilize *N*-phenylacetamide as a nitrogen source [100]. Besides a modification in the structure of the enzyme, these strains may also have mutated at the regulation gene level.
e) Mutants able to utilize phenylacetamide as a nitrogen source [94,105,106]. These mutants are able to hydrolyze phenylacetamide. However, some of these strains can no longer hydrolyze formamide and acetamide while others hydrolyze acetamide only very slowly.
f) Mutants able to utilize *N*-phenylacetamide and phenylacetamide as a nitrogen source [107]. These mutants were isolated from those already able to grow on *N*-phenylacetamide using it as their nitrogen source.

6.2.3 Purification and Characterization of the Amidase from *Pseudomonas aeruginosa*

The amidase of the wild type and of some mutants has been purified. The molecular weight of the amidase from the wild type is 200000. This enzyme is made up of 6 identical subunits [108].

The amino-acid compositions of the amidase from the wild type and from some mutants were studied. In the case of the enzyme from the strain able to utilize *N*-phenylacetamide, Brown and Clarke [100] showed that an isoleucine replaced a threonine. In the enzyme of the strain able to utilize butyramide, a serine had been replaced by a phenylalanine [109].

It requires the modification of only one amino acid to enable the enzyme to attack other substrates.

7 Analytical Techniques Used in the Study of Biological Hydrolysis of Amides

Amidases hydrolyze amides into organic acids in their ammonium salt form. In order to demonstrate or assay this enzymatic activity three approaches could be considered:

a) Measurement of the disappearance of the amide. This could be accomplished using gas chromatography or nuclear magnetic resonance [110–112].
b) Measurement of the appearance of the organic acid. This could be done by TLC.
c) Measurement of the appearance of the ammonium ion. This could be performed with colorimetric assay according to the method of Muftic [113].

7.1 Measurement of the Disappearance of the Amide

7.1.1 Gas Chromatographic Technique

Gas chromatography is a technique not often used in enzyme analysis. Di Geronimo and Antoine [74] and Mimura et al. [114] used this technique for the study of nitrile metabolism in a *Nocardia* and a *Corynebacterium*. The following compounds could be assayed by GLC: formamide, acetamide, propionamide, butyramide, isobutyramide, valeramide, isovaleramide, pivalamide, acrylamide, methacrylamide and crotonamide.

a) *Apparatus and operational conditions*

In our work, we used an Intersmat IGC 121 DFL gas chromatograph equipped with a flame ionization detector. This apparatus is connected to a ICR-1B calculator printer which prints out the chromatogram and the pertinent calculations. The column used was a 0.5-m long, 3-mm interior diameter nickel tube filled with Porapak Q (80–100 mesh). The temperature conditions were as follows: oven 170°–240 °C, injector 250 °C, detector 300 °C.

The temperature of the oven was set according to the amide to be assayed in order to obtain a complete analysis run within 5 minutes. The gas flows were as follows:
- vector gas: nitrogen 25 ml min^{-1} (0.4 bar)
- burning gas: hydrogen 25 ml min^{-1}.
  reconstituted air 400 ml min^{-1}

b) *Measurement of the amidase activity by this technique*

First, a standard curve was obtained using amide solutions at concentrations varying from 1 to 50 mM. An aliquot of 2 µl of each solution was injected into the column.

Enzyme kinetic experiments were performed at 30 °C and pH 7. A test tube containing a 2 ml mixture of phosphate buffer (50 mM; pH 7) and amide substrate was tempered in a water bath at 30 °C. Following 2–3 min of tempering, the reaction was started by adding 0.5 ml of the enzyme extract appropriately diluted in phosphate buffer. Aliquots of the reaction mixture were taken at regular time intervals with a chromatographic syringe and directly injected into the gas chromatograph, which instantly stopped the enzyme reaction. Figure 1 gives the chromatogram of the enzyme kinetics for the hydrolysis of acetamide by an enzyme extract of the wild type *Brevibacterium* sp. R312. The hydrolysis rate of the above-cited amides could thus be measured.

7.1.2 Nuclear Magnetic Resonance Method

a) *Apparatus*

The NMR analyses were performed on two types of apparatus. Most compounds were detected with a Varian EM 360 spectrometer (probe temperature 30 °C). Meth-

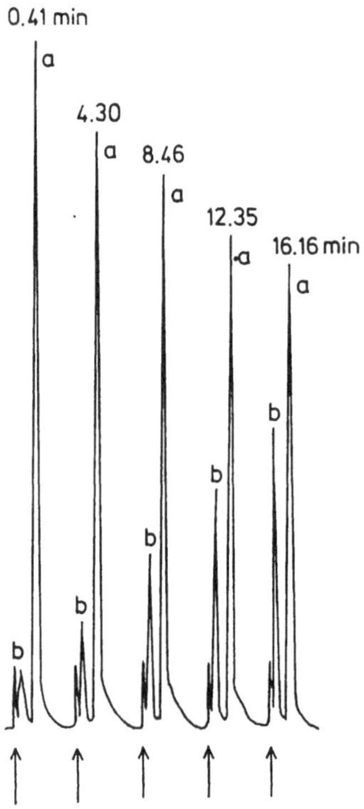

Fig. 1. Chromatography of the enzyme kinetics for the hydrolysis of acetamide by an enzyme extract of *Brevibacterium* sp. R312. ↑: injection; a: peak of acetamide; b: peak of acetic acid

acrylamide, acrylamide and their hydrolysis products were analyzed with a Varian HA 100 spectrometer. The external reference used for the determination of chemical shifts was a hexamethyldisiloxane solution.

b) *Chemical shift of amide and organic acid protons*

The assay of the amidase activity by the NMR technique could only be performed when the chemical shifts of some protons of the amide were sufficiently different from those of the corresponding acid in order to result in distinct peaks. Table 7 gives the proton chemical shift values of some amides and their corresponding acids.

c) *Measurement of the amidase activity by the NMR technique*

Except for propionamide and acrylamide, which gave complex multipeaks, and for α-aminopropionamide, the hydrolysis of the amides cited in Table 7 could be monitored by the NMR technique almost continuously. The reaction was started by the introduction of the enzyme extract (0.1 ml) into the NMR tube containing a mixture of substrate (0.1 ml) and buffer (0.3 ml). The reaction was monitored by recording NMR spectra at regular time intervals. An example of a reaction thus monitored is given in Fig. 2. The percentage hydrolysis of the amide could be determined from the height ratios of the substrate and the product. This corresponds to the concentration ratios of the compounds involved:

Table 7. Chemical shifts for different amides and acids (aqueous solutions at pH 7.0). Hexamethyldisiloxane was used as external standard

Compound	Protons examined	δ (Hz) Amide ($X = CONH_2$)	Acid ($X = COOH$)
CH_3-X	CH_3	70 (s)	67.5 (s)
CH_3-CH_2-X	CH_3	42.5 (t)	41.0 (t)
$(CH_3)_2CH-X$	CH_3	46.5 (d)	44.5 (d)
$(CH_3)_3C-X$	CH_3	46.0 (s)	43.5 (s)
$CH_2=CH-X$	$=CH-$	6.12 (q)	5.88 (q)
$CH_2=C(CH_3)-X$	CH_3	2.21 (s)	2.20 (s)
	CH_2	6.10 (s)	6.08 (s)
$CH_2(X)(OH)$	CH_2	136.0 (s)	131.5 (s)
$CH_3-CH(X)(OH)$	CH_3	54.0 (d)	52.0 (d)
$(CH_3)_2C(X)(OH)$	CH_3	53.0 (s)	51.0 (s)
$CH_2(X)(NH_2)$	CH_2	122.0 (s)	118.0 (s)
$CH_3-CH(X)(NH_2)$	CH_3	55.5 (d)	55.5 (d)
$CH_3-CH(X)(NH-CHO)$	CH_3	55.5 (d)	53.5 (d)
$CH_3-CH(X)(NH-CH_3)$	CH_3-C	52.75 (d)	56.0 (d)
$(CH_3)_2C(X)(NH_2)$	CH_3	59.0 (s)	52.0 (s)

s: singlet; d: doublet; t: triplet; q: quadruplet

$$\% \text{ hydrolysis} = \frac{h_c}{h_b + h_c} \times 100 ,$$

where h_b is the amide peak height and h_c is the acid peak height.

It is noteworthy that hydrolyzed amides give rise to acid compounds which are more or less protonated depending on the pH. This in turn gives a variable chemical shift of the protons close to the amide and acid function which depends on the pH. Also, for some pH values the resonance peaks overlap each other. In particular, the doublets of methyl groups of α-alanine and of α-aminopropionamide at pH 7.5 and 8 or the singlets of methyl group of acetamide and acetic acid at pH 3 and 4 overlap

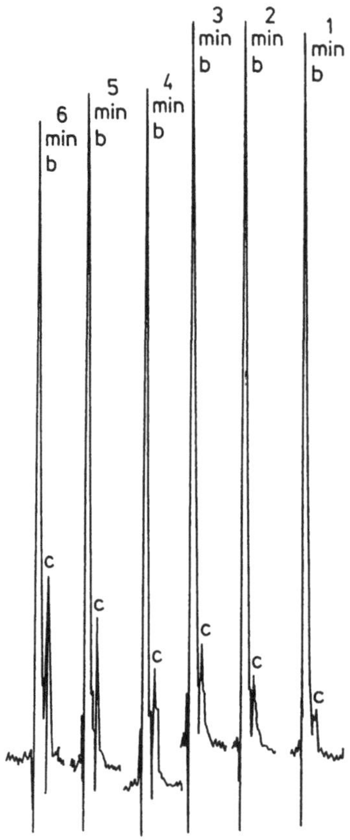

Fig. 2. Kinetics of hydrolysis of acetamide into acetic acid by strain *Brevibacterium* sp. R312 at 30 °C and pH 7.0 (phosphate buffer 0.1 M). Methyl peaks of acetamide (b) and acetic acid (c). The reaction was monitored continuously by recording a spectrum every minute

each other. In such cases, the enzyme kinetics could be monitored discontinuously. The reactions were performed in thermostated test tubes in buffered solutions. Aliquots were taken every minute and the reaction was stopped by the addition of one drop of concentrated HCl. Following pH adjustment, NMR analysis could then be performed.

The pH of the mixture may also vary during the hydrolysis reaction as the compounds involved are concentrated (up to 10^{-1} M) and have very different K_A acidity constants. This problem is particularly acute for the hydrolysis of α-aminoamides, which liberates α-amino acids and ammonia. It was necessary to ensure that for all hydrolysis reaction to be studied the buffer used had a high enough ionic strength to prevent any pH shift.

In conclusion, the above results show that the NMR technique can detect the amidase activity on several substrates and that the corresponding hydrolysis kinetics can be monitored by this method. The technique is easy to operate since the measurements are made with a single NMR tube. Sampling errors are avoided. Enzyme hydrolysis kinetics at different temperatures could eventually be monitored with a NMR apparatus equipped with a thermostatic control for the probe.

However, the NMR technique can only be used for a limited number of amides and at concentrations relatively high compared to those usually seen with classical enzyme analysis techniques. Also, the K_m of the amidases cannot be obtained by this method. The amounts of enzyme extract required for the test are also relatively large.

In any case, the NMR technique remains useful for the detection and study of a number of amidase activities difficult to monitor by other techniques.

7.2 Measurement of the Appearance of Organic Acid by TLC

The α-aminoamides, β-aminopropionamide, nicotinamide, isonicotinamide and their corresponding organic acids are not sufficiently volatile in order to be assayed by GLC. However, these compounds could easily be detected by TLC. This technique is difficult to apply quantitatively and was most often used to determine qualitatively whether a particular amide was a substrate of the amidase being studied.

7.2.1 Solvent Systems

Two solvent systems were used:
a) Propanol/ammonia 20% (70/30 v/v). This solvent was used for α-aminoamides and β-aminopropionamide.
b) Ethanol/chloroform/ammonia 20% (59/25/16 v/v/v). This solvent was used for nicotinamide and isonicotinamide.

7.2.2 Support

The support was a 0.1-mm thick layer of silica gel (60 F 254 Merck).

7.2.3 Sample Spotting

The samples were spotted 1.5 cm from the edge of the plate. Up to 19 samples could be spotted on a 20 × 20 cm plate. Aliquots of 1 μl containing 1–5 μg of the compounds to be assayed were spotted.

7.2.4 Chromatography Run

A solvent front travel of about 10 cm is usually sufficient.

7.2.5 Visualization

The plates were dried for 10 min at 110 °C, and sprayed with ninhydrin (2 g l^{-1} ninhydrin in a ethanol/acetic acid; 80/20 v/v mixture) for the detection of α-aminoacids, β-aminopropionamide and their corresponding organic acids. The TLC plates were then held over a hot plate until the colored spots appeared.

For the detection of nicotinamide, isonicotinamide and their corresponding organic acids, the plates were studied under UV light (254 nm) following drying.

7.2.6 Detection of Amidase Activity by TLC

The enzyme hydrolysis kinetics were run as described in Sect. 7.1.1. During the reaction, three or four samples were taken and spotted on the TLC plate. Two controls were also spotted, one for the substrate and the other for the hydrolysis product.

7.3 Measurement of the Appearance of the Ammonium Ion

Some amides could not be detected by GLC and only with difficulty with NMR techniques. Also, their corresponding organic acids could not be detected by TLC. The amidase activity on these substrates could only be detected by the assay of ammonium ions evolved in the reaction mixture. In this case, the colorimetric assay of ammonia developed by Muftic [113] was used.

The presence of amino groups of the enzyme extract in the reaction mixture would interfere with the assay of ammonia by this method. This assay must be preceded by the application of the Conway [115,116] microdiffusion step.

7.3.1 Conway Microdiffusion Method

The addition of a solution of potassium carbonate to an ammonia solution liberates gaseous ammonia which can then be trapped by a boric acid solution. This process takes place in a Conway cell, which has two circular compartments as shown in Fig. 3.

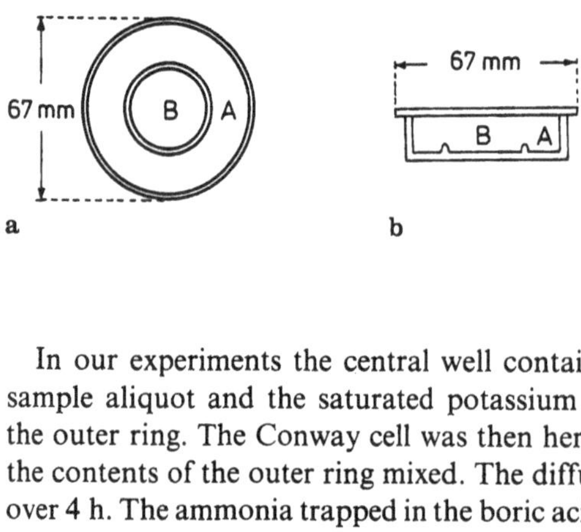

Fig. 3a and b. Schematic of a Conway cell

In our experiments the central well contained the boric acid solution (1%). The sample aliquot and the saturated potassium carbonate solution were pipetted into the outer ring. The Conway cell was then hermetically closed with a glass cover and the contents of the outer ring mixed. The diffusion process was allowed to take place over 4 h. The ammonia trapped in the boric acid could then be assayed by colorimetry. A control was required, in which the sample was replaced by distilled water, for the next step of the assay.

7.3.2 Colorimetric Assay

A mixture of phenol-hypochlorite-ammonia produces a blue coloration due to an indophenol polymer. Two reagent solutions were used. Solution 1 consisted of phenol (20 g) dissolved in 95% ethanol (100 ml). Solution 2 was made by dissolving 25 g calcium hypochlorite in 300 ml hot distilled water. Then 135 ml of a K_2CO_3 solution (200 g l^{-1}) were added to this mixture, which was heated and adjusted to 500 ml. The assay was performed by adding 0.5 ml of solution 1 then 0.1 ml of solution 2 to 1 ml of sample. After 5 min color development, the mixture color intensity was measured at 655 nm in a Beckman spectrophotometer. A standard curve was previously obtained with a solution of $(NH_4)_2SO_4$.

7.3.3 Assay of Enzyme Activity by the Colorimetric Technique

The enzyme reaction was performed as described in Sect. 7.1.1. It was necessary to use a larger reaction mixture volume (5–10 ml). At regular intervals, 1 ml aliquots were taken and immediately pipetted into the Conway cells. Mixing with the potassium carbonate immediately stopped the enzyme reaction.

8 The Amidases of Brevibacterium sp. R312

8.1 Isolation and Selection of the Wild Type *Brevibacterium* sp. R312

The selection was performed from samples of soil where nitrile compounds were known to be present. Following dilution in saline water, the soil was spread plated onto a solid medium containing Difco-Yeast Carbon Base (1.17%), acetonitrile (0.1%), and agar (2.5%). The colonies which appeared were purified and tested again in a liquid medium with acetonitrile as the sole nitrogen source. The same medium without acetonitrile was used as a control. Only the strains able to grow with acetonitrile as their sole nitrogen source following four or five transfers were kept. No yeast or mold was obtained. Only bacteria were selected. The strains isolated were identified [117,118] and were found to belong to several bacterial genera: *Bacillus*, *Bacteridium*, *Micrococcus*, and *Brevibacterium*. Among these, the *Brevibacterium* sp. R312 was further studied since it presented several features interesting for an eventual industrial application (flocculation, growth rate, etc.). This strain is able to transform acetonitrile into ammonium acetate via an acetamide intermediary state. The enzyme involved in the hydratation of acetonitrile into acetamide was shown to be a single system. This enzyme was named "nitrile hydratase" and has a large activity spectrum [119].

The *Brevibacterium* sp. R312 is able to hydrolyze acetamide. It was thought it would be interesting to determine the activity spectrum of this strain, and if the spectrum were wide, to determine whether a single or several enzymes were involved in this hydrolysis reaction.

8.2 The Amidase Activity Spectrum of the *Brevibacterium* sp. R312 Strain

All the amides mentioned in Table 8 were hydrolyzed by the *Brevibacterium* sp. R312 strain. Only the internal amides (lactams), ε-caprolactam and α-amino-ε-caprolactam, and the amide analogues (thioacetamide, *N*-methylacetamide, acetohydroxamic acid, etc.) were not attacked by this strain.

8.3 Detection of an Amidase with a Wide Activity Spectrum and of a L-α-Aminoamidase

Several mutant strains unable to hydrolyze fluoroacetamide were selected from the wild type *Brevibacterium* sp. R312. The selection technique was based on the toxicity of fluoroacetic acid for microorganisms [84,87,104]. The principle of the technique, inspired

Table 8. Amides tested as substrates of the amidase

Amide	Chemical formula	Analytical method
Formamide	$H{-}CONH_2$	col. NH_4^+
Acetamide	$CH_3{-}CONH_2$	GC
Propionamide	$CH_3{-}CH_2{-}CONH_2$	GC
Butyramide	$CH_3{-}CH_2{-}CH_2{-}CONH_2$	GC
Valeramide	$CH_3{-}CH_2{-}CH_2{-}CH_2{-}CONH_2$	GC
Isobutyramide	$(CH_3)_2CH{-}CONH_2$	GC
Pivalamide	$(CH_3)_3C{-}CONH_2$	GC
Acrylamide	$CH_2{=}CH{-}CONH_2$	GC
Methacrylamide	$CH_2{=}C(CH_3){-}CONH_2$	GC
Crotonamide	$CH_3{-}CH{=}CH{-}CONH_2$	GC
Vinylacetamide	$CH_2{=}CH{-}CH_2{-}CONH_2$	GC
Hydroxyacetamide	$HO{-}CH_2{-}CONH_2$	col. NH_4^+
Lactamide	$CH_3{-}CH(OH){-}CONH_2$	col. NH_4^+
α-Hydroxyisobutyramide	$(CH_3)_2C(OH){-}CONH_2$	col. NH_4^+
Urea	$NH_2{-}CONH_2$	col. NH_4^+
Benzamide	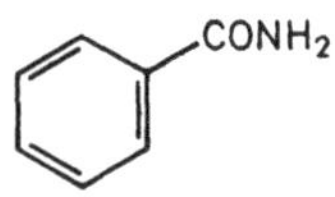	col. NH_4^+
Nicotinamide	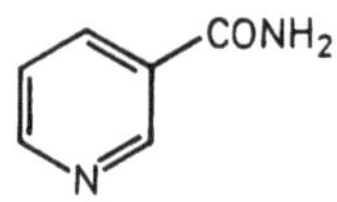	TLC
Isonicotinamide	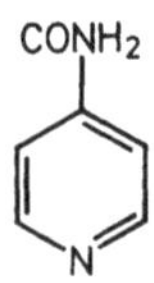	TLC
Fluoroacetamide	$F{-}CH_2{-}CONH_2$	col. NH_4^+
Glycinamide	$NH_2{-}CH_2{-}CONH_2$	TLC
Alanine amide (D)	$CH_3{-}CH(NH_2){-}CONH_2$	TLC
(DL)	$CH_3{-}CH(NH_2){-}CONH_2$	TLC
(L)	$CH_3{-}CH(NH_2){-}CONH_2$	TLC
α-Aminobutyramide (DL)	$CH_3{-}CH_2{-}CH(NH_2){-}CONH_2$	TLC
Valine amide (DL)	$(CH_3)_2CH{-}CH(NH_2){-}CONH_2$	TLC
Leucinamide (DL)	$(CH_3)_2CH{-}CH_2{-}CH(NH_2){-}CONH_2$	TLC
Isoleucinamide (DL)	$CH_3{-}CH_2{-}CH(CH_3){-}CH(NH_2){-}CONH_2$	TLC
Phenylalanine amide (DL)	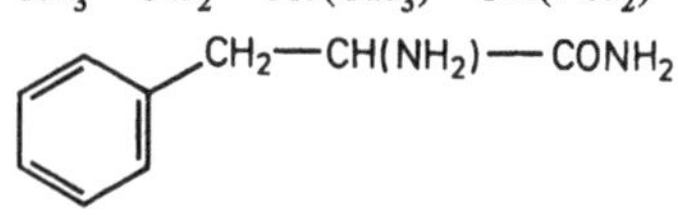	TLC
Methionine amide (DL)	$CH_3{-}S{-}(CH_2)_2{-}CH(NH_2){-}CONH_2$	TLC
N-formyl-α-aminopropionamide	$CH_3{-}CH(NH{-}CHO){-}CONH_2$	col. NH_4^+
β-Aminopropionamide	$NH_2{-}CH_2{-}CH_2{-}CONH_2$	TLC
N-Methyl-α-aminopropionamide	$CH_3{-}CH(NH{-}CH_3){-}CONH_2$	col. NH_4^+
Malonamide	$H_2NCO{-}CH_2{-}CONH_2$	col. NH_4^+
Succinamide	$H_2NCO{-}CH_2{-}CH_2{-}CONH_2$	col. NH_4^+
Adipamide	$H_2NCO{-}CH_2{-}CH_2{-}CH_2{-}CH_2{-}CONH_2$	col. NH_4^+

col. NH_4^+ : colorimetric assay of ammonia
GC: gas chromatography
TLC: thin layer chromatography

Table 9. Principle of selection of mutants unable to hydrolyze fluoroacetamide

Medium	Wild type R 312	Mutant
Minimum Medium (MM) + Ammonium acetate	Growth	Growth
Minimum Medium (MM) + Ammonium acetate + Fluoroacetamide	No growth, $FCH_2{-}CONH_2 \rightarrow FCH_2{-}COOH$ (toxic)	Growth
Minimum Medium (MM) + Acetamide	Growth $CH_3{-}CONH_2 \rightarrow CH_3{-}COONH_4$	No growth (no carbon or nitrogen source)

The composition of the basal Minimum Medium (MM) has been given previously [120]

by that of Clarke and Tata [104], is shown in Table 9. The cells of the wild type are able to hydrolyze fluoroacetamide into toxic fluoroacetic acid and could not develop. The mutant strains unable to hydrolyze fluoroacetamide could grow on the selective medium.

Several mutants were isolated with and without mutagenesis with EMS. These strains were unable to hydrolyze acetamide and fluoroacetamide. Only the *Brevibacterium* sp. A4 mutant was studied further. This strain has the same nitrile hydratase activity spectrum as the wild type. However, this strain was shown to be unable to hydrolyze many amides (Table 10). Only formamide, urea, nicotinamide, and α-aminoamides were hydrolyzed by this mutant strain. Also, only the L forms of the α-aminoamides were hydrolyzed into L-α-amino acids (Table 11).

Several enzyme systems seemed to be responsible for the hydrolysis of amides. The wild type *Brevibacterium* sp. R312 was thus shown to have an amidase with a large activity spectrum which only attacks compounds with a true amide function, a L-α-aminoamidase which hydrolyzes L-α-aminoamides into L-α-aminoacids and several enzymes specific for urea, formamide, L-glutamine and nicotinamide. Only the first two enzymes were studied further.

9 The L-α-Aminoamidase

The purification of the particulate L-α-aminoamidase [127] that is associated with the solid fractions of the cells was achieved according to a method derived from that used by Schneider et al. [128] for the particulate adenylate cyclase from *Nocardia restricta*.

The L-α-aminoamidase from *Brevibacterium* sp. A4 was found to have a molecular weight of 135000. This enzyme activity was maximum for pH values between 9 and 9.5 and was nil below 5 and above 12.

The activation energy for the enzyme reaction with DL-α-aminopropionamide as

Table 10. Amidase activities lost by the mutant strain *Brevibacterium* sp. A4

Compounds	Chemical formula	R or n	Whole cells		Sonicated cell suspension (S_1)		Analytical methods
			Strain R312	Strain A4	Strain R312	Strain A4	
Aliphatic amides	$R-C(=O)-NH_2$	CH_3, C_2H_5 $(CH_3)_2CH$, $(CH_3)_3C$	+	—	+	—	NMR; GC
Aliphatic diamides	$(CH_2)_n(C(=O)NH_2)_2$	1, 2, 4	+	—	+	—	E_{NH_3}; col. NH_4^+
α-Unsaturated amides	$CH_2=C(R)-CO-NH_2$	H, CH_3	+	—	+	—	GC
β-Unsaturated amides	$CH_3-CH=CH-C(=O)NH_2$		+	—	+	—	GC; E_{NH_3}; col. NH_4^+
	$CH_2=CH-CH_2-C(=O)NH_2$		+	—	+	—	GC; E_{NH_3}; col. NH_4^+
α-Benzenic amide	$\Phi-CO-NH_2$		+	—	+	—	E_{NH_3}; col. NH_4^+
D-α-Aminopropionamide	$CH_3-CH(NH_2)-CONH_2$ (D)		+	—	+	—	TLC; NMR
N-Substituted α-aminoamides	$CH_3-CH(NHR)-CONH_2$	CH_3, CHO	+	—	+	—	NMR; col. NH_4^+

α-Hydroxyamides	$RCH(OH)CONH_2$	CH_3, $(CH_3)_2$	+	—	+	—	NMR; col. NH_4^+
β-Aminopropionamide	$NH_2-(CH_2)_2-CONH_2$		+	—	+	—	TLC
Heterocyclic amides	$R-CO-NH_2$	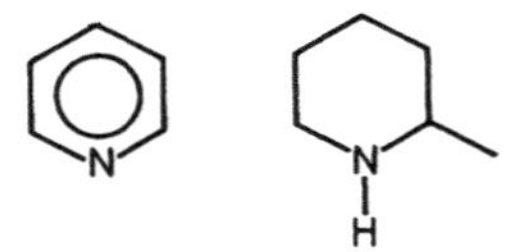	+	—	+	—	TLC
Heterocyclic diamide	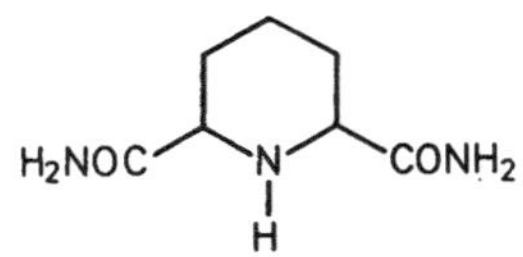		+	—	+	—	TLC

NMR: nuclear magnetic resonance [111, 112]
TLC: thin layer chromatography [72]
GC: gas chromatography [110, 112]
col. NH_4^+: colorimetric assay of the ammonium ion [113]
E_{NH_3}: ammonia-specific electrode [121–126]

Table 11. Amidase activities of the acetamidase-defective mutant strain *Brevibacterium* sp. A4

Compounds	Chemical formula	R	Whole cells		Sonicated cell suspension (S_1)		Analytical method
			Strain R312	Strain A4	Strain R312	Strain A4	
Formamide	$HCONH_2$		+	+	+	+	GC; E_{NH_3}; col. NH_4^+
Urea	H_2NCONH_2		+	+	+	+	E_{NH_3}; col. NH_4^+
Nicotinamide	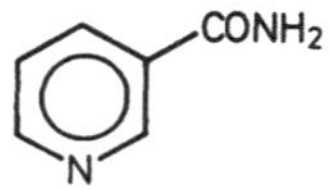		+	+	+	+	TLC
L-Glutamine	$H_2NCOCH_2CH_2CH<^{COOH}_{NH_2}$ (L)		—	—	+	+	TLC
Glycinamide	$H_2N-CH_2CO-NH_2$		+	+	+	+	TLC
DL-α-Aminoamides	$R-CH<^{CONH_2}_{NH_2}$ (DL)	CH_3, C_2H_5, $(CH_3)_2CH$ $CH_3SCH_2CH_2$, ΦCH_2	+	50%	+	50%	TLC
L-α-Aminopropionamide	$CH_3CH<^{CONH_2}_{NH_2}$ (L)		+	+	+	+	TLC; NMR

NMR: nuclear magnetic resonance [111, 112]
TLC: thin layer chromatography [72]
GC: gas chromatography [110, 112]
col. NH_4^+: colorimetric assay of the ammonium ion [113]
E_{NH_3}: Ammonia-specific electrode [121–126]

Table 12. K_m and relative rates of hydrolysis of some α-aminoamides and of glutamine. A 100 hydrolysis rate was chosen for L-α-aminopropionamide ($V_{max} = 0.16 \times 10^{-6}$ mole min^{-1} mg^{-1} protein). All the substrates have the formula $R{-}CH\langle^{CONH_2}_{NH_2}$

Substrate	K_m (10^{-3} M)	R	Relative hydrolysis rate
L-Serinamide	6	HOH_2C-	10
Glycinamide	9	$H-$	15
L-Tryptophanamide	5	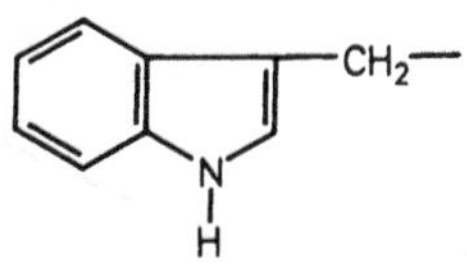	25
L-Phenylglycinamide	10	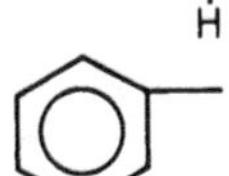	80
L-Glutamine	10	$NH_2-C(=O)-CH_2-CH_2-$	90
L-Aminopropionamide	9	H_3C-	100
L-Phenylalaninamide	8	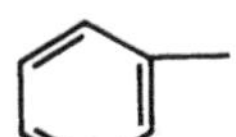	150
L-Valinamide	8	$(H_3C)_2CH-$	165
L-Isoleucinamide	11	$(H_3C-CH_2)(H_3C)CH-$	200
L-Methioninamide	9	$H_3C-S-H_2C-H_2C-$	260
L-Amino 2-butyramide	8	CH_3-CH_2-	325
L-Leucinamide	6	$(H_3C)_2CH-CH_2-$	2150

substrate was calculated between 20° and 60 °C and then was estimated to be 6.3 kcal per mole. Calcium was found to be inhibitory at 5×10^{-3} M and above. Hg^{2+}, Co^{2+} and Zn^{2+} were inhibitory at 10^{-4} M. No cation was found to activate the non-dialyzed enzyme preparation. Phosphate salts were shown to have a repressing effect on the biosynthesis of this enzyme. However, the enzyme biosynthesis does not depend on the nature of the carbon and nitrogen source.

The values of K_m and V_{max} were determined for different substrates. The K_m were not significantly different, but V_{max} greatly varied depending on the substrate tested (Table 12). Figure 4 shows the logarithm of the relative rate of hydrolysis of different substrates versus the mass of the R radical attached to the carbon of each amide. Several conclusions can be made from this representation:

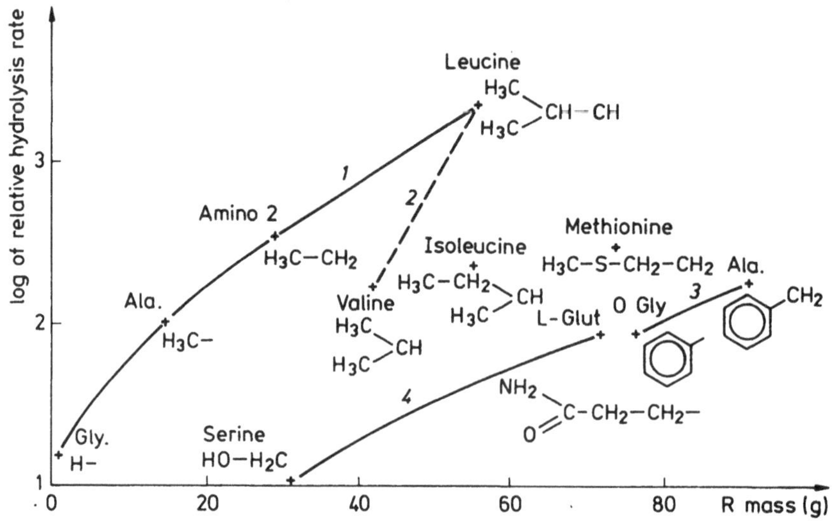

Fig. 4. Logarithm of the relative hydrolysis rate versus the mass of the R radical of each amide

a) A lengthening of the R radical results in a greater rate of hydrolysis of the substrate by this enzyme (curve 1). A similar result was observed on a leucine aminopeptidase by Smith and Hill [8].

b) Leucinamide shows a greater hydrolysis rate than valinamide. In this case, there is also a lengthening of the R radical (curve 2). It is noteworthy that isoleucinamide is less rapidly hydrolyzed than leucinamide, but as rapidly hydrolyzed as valinamide. This result would mean that the larger size of the

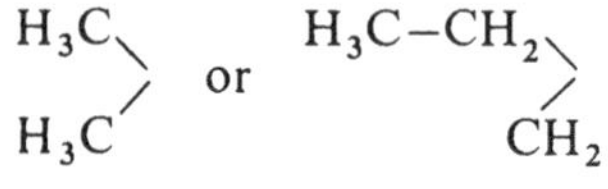

branchings would facilitate the hydrolysis of the substrate.

c) The larger size of the benzene ring also increased the relative hydrolysis rate (curve 3).

d) Finally, the immediate proximity of an oxygen atom decreases the hydrolysis rate of the substrate (curve 4).

The results for methioninamide and tryptophanamide are less easily interpreted.

The enzyme studied above is specific for L-α-aminoamides. A single enzyme system hydrolyzes all the L-α-aminoamides. This enzyme was also shown to be able to hydrolyze dipeptides (L-alanyl-L-alanine). This is a membrane-associated enzyme that is resistant to high temperatures, especially when non-purified. Its general properties are close to those of many α-aminopeptidases previously described in the literature [25–29, 32, 43–49, 52–54, 129–133].

10 The Amidase with a Wide Activity Spectrum

10.1 Properties of the Enzyme

The wide activity amidase is a soluble enzyme, as previously confirmed by Miller and Knowles [134]. It was found to have a molecular weight of 180000 [135]. The molecular weight of this enzyme is close to that found for the amidase of *Pseudomonas aeruginosa* (200000) by Brown et al. [136]. On the other hand, the molecular weight of the amidase from *Arthrobacter* sp. J-1 was previously found to be much larger [76] (320000).

The method of Laemmli [137] was used to study the multimer composition of the amidase from *Brevibacterium* sp. R312. The electrophoretic analysis shows that the amidase is composed of a single type of subunit. Each subunit has a molecular weight of 43000 ± 2000). The quaternary structure of the amidase from *Brevibacterium* sp. R312 is thus very different from that of *Pseudomonas aeruginosa* which has six subunits of about 33000 molecular weight [136] and that of *Arthrobacter* J-1 which has eight units of about 40000 Da [76].

The isoelectric pH was estimated by electrofocusing to be 3.5. The result obtained was very close to that obtained for the amidase of *Arthrobacter* sp. J-1 (pI 3.8) [76].

The enzyme activity was maximum for pH values between 6 and 7. The activation energy of the reaction between 20 and 60 °C was calculated to be 8.6 kcal per mole.

The amidase possesses at least one —SH group in its active site. First, the inhibition by iodoacetic acid and *N*-ethylmaleimide indicated that an —SH group could be involved. This hypothesis was later confirmed by the total inactivation of the enzyme in the presence of parahydroxymercuribenzoate.

This enzyme was shown by genetic studies to have a large spectrum. It could hydrolyze aromatic as well as aliphatic amides. The range of compounds hydrolyzed is much more important than those previously described in the literature for other amidases [56,60,67–69,76,93,94].

The K_m and V_{max} of the amidase were determined for 11 different substrates (Table 13). The K_m results for the n aliphatic amide series (acetamide, propionamide, butyramide, valeramide) showed that this enzyme had the weakest affinity for propionamide (K_m = 44 mM). The affinity for acetamide was 13–20 times greater. Also, the precision of the results could not show any clear difference between the K_m of butyramide and valeramide. The averages of the results obtained seemed to indicate that the affinity of the enzyme for a substrate increased with a longer carbon radical. Except for acetamide compared to propionamide, the results of the n aliphatic amides and of isobutyramide, pivalamide and benzamide indicated that a longer and larger carbon chain of the substrates increased the affinity of the enzyme. The affinity of the amidase increased with more bonding possibilities. On the other hand, considering K_m values for acetamide and isobutyramide compared to those of respectively propionamide and pivalamide, the spatial crowding of another methyl group must be responsible for the reduction of enzyme affinity for the two latter substrates, notwithstanding the increase in bonding possibilities.

The double bond of acrylamide modifies the spatial arrangement of this molecule compared to propionamide, D-α-aminopropionamide and β-aminopropionamide. Its spatial crowding for the active site of the enzyme was less, hence its smaller K_m value compared to those of the above substrates.

Table 13. K_m and V_{max} values for different substrates hydrolyzed by the wide spectrum amidase of *Brevibacterium* sp. R 312

Substrate		K_m (10^{-3} M)	Relative specific activity
C_2	Acetamide	3.5 2.6 2.1	9
C_3	Propionamide	49 38	100
	Acrylamide	16 8.4 11 15	21
	D-α-Aminopropionamide	30 30	2.5
	β-Aminopropionamide	20	0.2
	DL-Lactamide	10 6.0	16
C_4	Butyramide	0.28 0.55	1.4
	Isobutyramide	0.50 0.56	9
C_5	Valeramide	0.13 0.34	0.3
	Pivalamide	1.2 1.1 1.0	0.6
	Benzamide	3.0 2.5	0.8

The situation for benzamide is similar to that of acrylamide. This molecule should have had an important crowding effect due to its aromatic ring and a weak affinity by the enzyme was expected for this substrate. However, its affinity (K_m = 2.7 mM) was very close to the value measured for acetamide. This must be due to the fact that the benzene ring is flat so that the crowding effect was much less than expected, resulting in closer contact with the enzyme's active site.

DL-Lactamide showed an average K_m slightly smaller than that of acrylamide, and significantly smaller than that of D-α-aminopropionamide, although its size and spatial conformation are similar to this compound. The greater affinity of the amidase for the DL-lactamide molecule must be due to the presence of the hydroxyl group on the α carbon atom, which is available for establishing hydrogen bonds between the substrate and enzyme.

The D-α-aminopropionamide was chosen as the substrate in this study rather of a DL-α-aminopropionamide mixture because otherwise the results would have been difficult to discuss since the enzyme preparation also contained small amounts of a L-α-aminoamidase; it should be pointed out that the L-α-aminoamidase specificity for L-α-aminopropionamide ($K_m = 9$ mM) only represents at most about 2% of the general amidase activity. The presence of an amino group on the α carbon increased the enzyme activity for this substrate as compared to propionamide. The same amino group at a position further from the amide end, as it is the case with β-aminopropionamide, further increased the affinity of the enzyme. This must be due to the relative spatial effect of these two positions.

The greatest V_{max} was obtained for propionamide, which was chosen as the reference substrate for further experiments. The hydrolysis rate was smaller for aliphatic amides with a longer (butyramide, valeramide) or a shorter (acetamide) carbon chain than propionamide. The above results are similar to those reported for the amidases from *Pseudomonas aeruginosa*[67,102] and from *Pseudomonas fluorescens*[68].

The same decrease in hydrolysis rate was observed for substrates with increasing spatial crowding with $-CH_3$ groups (propionamide, isobutyramide, pivalamide) or with an aromatic ring (benzamide). It was noteworthy that the $-CH_3$ group crowding has a smaller reducing effect on the hydrolysis rate than the lengthening of the hydrocarbon chain.

The double bond of the acrylamide molecule had an adverse effect on the rate of enzymatic hydrolysis since the relative specific rate for this substrate represents only 20% of the rate obtained for propionamide.

Finally, when $-OH$ or $-NH_2$ groups were substituted onto the α position on the propionamide molecule, the enzymatic activity was reduced but remained relatively important (propionamide = 100, lactamide = 16, D-α-aminopropionamide = 2.5). On the other hand, when the amino group was substituted onto the β position, the amidase activity was barely detectable (β-aminopropionamide = 0.2).

The purified amidase of *Brevibacterium* sp. R312 was shown to be able to convert acetamide, acetate, and ethyl acetate into acetohydroxamate. The reactions catalyzed by this enzyme could be summarized as follows:

a) Amide hydrolase activity

$$\underset{\text{amide}}{R{-}C({=}O)NH_2} + H_2O \rightarrow \underset{\text{organic acid}}{R{-}C({=}O)OH} + NH_3$$

b) Amide transferase activity

$$\underset{\text{amide}}{R{-}C({=}O)NH_2} + \underset{\text{hydroxylamine}}{NH_2OH} \rightarrow \underset{\text{hydroxamic acid}}{R{-}C({=}O)NHOH} + NH_3$$

c) Acid transferase activity

$$R{-}C{\Large\substack{\nearrow O\\ \searrow OH}} + NH_2OH \rightarrow R{-}C{\Large\substack{\nearrow O\\ \searrow NHOH}} + H_2O$$

organic acid hydroxylamine hydroxamic acid

d) Ester transferase activity

$$R{-}C{\Large\substack{\nearrow O\\ \searrow OR'}} + NH_2OH \rightarrow R{-}C{\Large\substack{\nearrow O\\ \searrow NHOH}} + R'OH$$

ester hydroxylamine hydroxamic acid alcohol

While the amide hydrolase and amide transferase activity are due to two different enzymes in *Mycobacterium smegmatis* [64], many amide hydrolases are also able to transfer the acyl of the amide onto hydroxylamine forming hydroxamates [56,73,76,77]. The specific relative rates of these two reaction types were compared (Table 14). The transfer reaction of the acyl of acetamide onto hydroxylamine was about 30 times

Table 14. Comparison of the relative rates of the hydrolysis and transfer reactions

Reaction	Specific relative rate
$CH_3{-}CONH_2 + H_2O \rightarrow CH_3COOH + NH_3$	30
$CH_3{-}CONH_2 + NH_2OH \rightarrow CH_3{-}CONHOH + NH_3$	1000
$CH_3{-}COOH + NH_2OH \rightarrow CH_3{-}CONHOH + H_2O$	89
$CH_3COOCH_2CH_3 + NH_2OH \rightarrow CH_3{-}CONHOH + CH_3CH_2OH$	44

Table 15. Transferase activity of the amidase from *Brevibacterium* sp. R 312. Determination of K_m and V_{max} for different substrates

Substrate	Relative V_{max}	K_m (mM)
Acetamide	1000	31
Propionamide	280	88
Acrylamide	490	93
Acetate	89	83
Propionate	1113	628
Acrylate	33	88
Ethyl acetate	44	18
Ethyl propionate	135	16
Ethyl acrylate	147	218
Acetamidine	0	—
N-Methylacetamide	5.6	62
N,N'-Dimethylacetamide	0	—

more rapid than the hydrolysis of acetamide into ammonium acetate. Similarly, the acetamide transfer reaction was respectively about 10 and 20 times faster than the acetate and the acetyl-acetate transfer reaction onto hydroxylamine (Table 15). Thus, in the presence of hydroxylamine, the enzyme transforms acetamide into acetohydroxamic acid rather than into ammonium acetate [138].

Among amides, acetamide appeared to be the substrate most quickly transformed by the enzyme. It should also be noted that the order of the relative rates of amide hydrolysis (propionamide > acrylamide > acetamide) was the reverse of the order of transfer rates of the acyl group onto hydroxylamine (acetamide > acrylamide > propionamide). These results are identical to those previously observed for the amidase of *Pseudomonas aeruginosa* [56].

Concerning the organic acids, propionate was the substrate most rapidly transformed, the amidase has a very low affinity for this compound (K_m = 628 mM). The enzyme from *Pseudomonas aeruginosa* showed the greatest rate of transfer for acetate among the acids tested [56].

Ethyl acrylate was the ester most rapidly attacked. It is also the substrate for which the enzyme has the least affinity (K_m = 218 mM). These results are clearly distinct from those obtained for the amidase of *Pseudomonas aeruginosa* which transferred ethyl acetate most quickly [56].

Among the other compounds tested, only *N*-methylacetamide was converted into acetohydroxamic acid by the amidase. The reaction rate for this compound was, however, 180 times slower than that obtained for the transfer of the acyl group of acetamide onto the hydroxylamine. Also, the affinity of the amidase for acetamide (K_m = 31 mM) was greater than that determined for *N*-methylacetamide (K_m = 62 mM).

No hydrolysis of *N*-methylacetamide by the amidase of *Brevibacterium* sp. R312 could be detected. It was thus surprising to find that in the presence of hydroxylamine, this amide analogue was converted by the enzyme into acetohydroxamic acid. This result led to the supposition that hydrolysis of *N*-methylacetamide did occur, however at such a slow rate that it was not detected by GLC monitoring [138].

10.2 Mechanism of Reactions Catalyzed by the Amidase with a Wide Spectrum

The transfer of the acyl group of acetamide onto hydroxylamine is a "Bibi Ping-Pong" mechanism reaction [139]. The reaction occurs according to the following scheme:

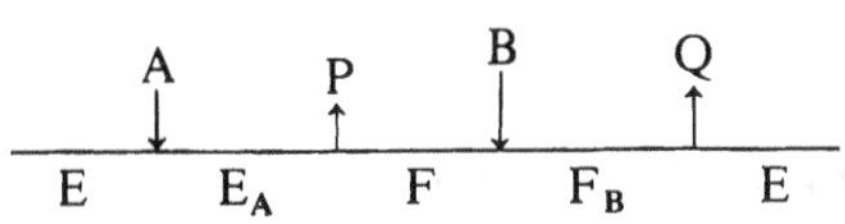

where A represents acetamide, P the ammonium ion, B hydroxylamine, Q acetohydroxamic acid, and E, EA, F, and FB represent the different forms taken by the enzyme during the reaction. In the case of the hydrolysis of acetamide, it is probable that the reaction mechanism is the same. Then A would represent acetamide, P the ammonium ion, B water, and Q acetic acid.

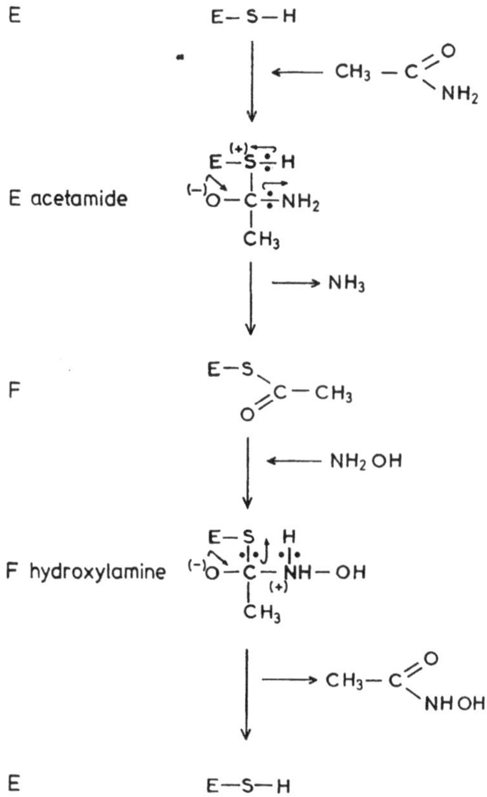

Fig. 5. Mechanism of the transfer reaction of the acyl group from acetamide onto hydroxylamine

The wide spectrum amidase of *Brevibacterium* sp. R312 possesses at least one —SH group. The mechanism of the two types of reaction catalyzed by this enzyme could thus be described as in Figs. 5 and 6. The enzyme fastens itself to acetamide and becomes an E-S-acyl intermediary with the departure of the first product NH_3. Hydroxylamine (transfer reaction) or water (hydrolysis reaction) then reacts with the intermediary yielding the second product (acetohydroxamic acid or acetic acid). Following the departure of the second product, the enzyme returns to its original state.

10.3 Regulation of the Wide Spectrum Amidase

The amidase from *Brevibacterium* sp. R312 is an adaptative enzyme. Only acetamide and *N*-methylacetamide can induce its biosynthesis [140,141].

The enzymes responsible for the catabolism of some carbon compounds in micro-organisms are often subject to the glucose effect. This phenomenon is referred to as catabolic repression [142] or metabolic repression [143]. The best-known example is that of β-galactosidase of *E. coli*. The biosynthesis of this enzyme is repressed when glucose is present in the culture medium while there is no repression in the presence of glycerol [142].

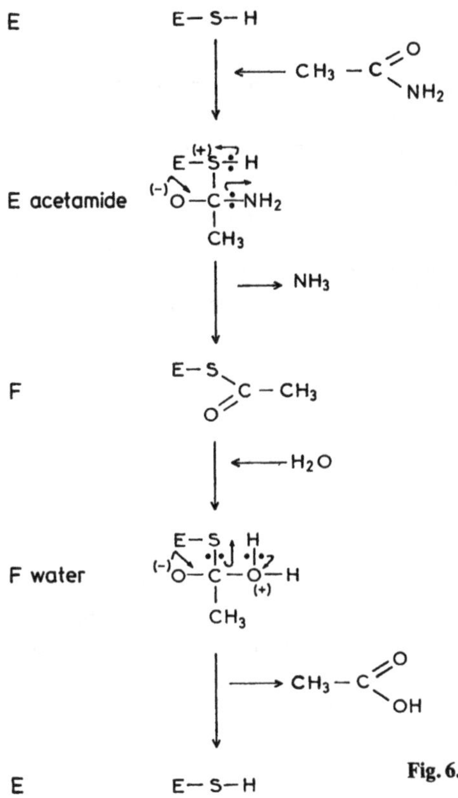

Fig. 6. Mechanism of the hydrolysis reaction of acetamide

Similarly, enzymes responsible for the catabolism of some nitrogen compounds are often repressed by the ammonium ion. This is the case for *Aspergillus nidulans* [144] and *Neurospora crassa* [145] nitrate reductases.

Some compounds such as amides may be utilized as either a carbon or a nitrogen source. The regulation of the biosynthesis of enzymes responsible for the catabolism of this type of product must thus be adaptative. If there is a catabolic repression, the compound (amide) may no longer be used as a nitrogen source. Similarly, if repression by the ammonium ion occurs, the compound may no longer be used as a carbon source.

The acetamidase of *Aspergillus nidulans* is inducible and its biosynthesis is only repressed when both glucose and ammonium sulfate are present in the culture medium. This enzyme is only synthesized when absolutely necessary.

The amidase from *Brevibacterium* sp. R312 was able to hydrolyze a large number of amides so that this bacteria could utilize some of these compounds (acetamide, propionamide, butyramide) as its sole nitrogen and carbon source. Neither glucose nor ammonium sulfate appeared to have any effect on the biosynthesis of this enzyme. The amidase was biosynthesized even if it was not required for the growth of the strain.

The above phenomenon could be explained in a different way. Besides its amidases, *Brevibacterium* sp. R312 happens to possess a nitrile hydratase with a wide substrate spectrum [119]. The biosynthesis of this nitrile hydratase is, similarly to the amidase, not subject to repression by glucose or by the ammonium ion [146,147]. When these compounds are present in the medium, these two enzymes are still synthesized so that nitriles are hydrolyzed into less toxic organic acids. Nitriles are very toxic, some are even mutagenic [148], and they could prevent the strain from growing. Thanks to this metabolic pathway, this strain could defend itself against a hostile environment by hydrolyzing toxic nitriles into assimilable products.

The amidase biosynthesis was repressed by all the organic acids. This phenomenon is difficult to interpret. It is probably not a simple feedback regulation phenomenon (enzyme biosynthesis repression by the reaction product) since the induction by *N*-methylacetamide was repressed by all the acids tested. It is more probable that the organic acids acted with their —COOH function, which is analogue to the $—CONH_2$ amide function, and competed with the inducer [140].

11 Applications

11.1 Production of L-α-Aminoacids

The industrial production of some α-aminoacids is mainly based on enzymatic processes: splitting of DL-*N*-acetyl-α-aminoacids with L-*N*-α-acylases, transamination of α-ketoacids, hydrolysis of racemic hydantoins by hydantoinases, addition of ammonia on fumaric acid catalyzed by an aspartase, decarboxylation of L-aspartic acid by an aspartate decarboxylase, production of L-citrulline by hydrolysis of L-arginine, etc. Patents for the production of L or D-α-aminoacids by biotransformation of racemic α-aminoamides using L-α-aminoacylamidase enzyme preparations have been taken out [149,150]. The α-aminoamides and their precursors α-aminonitriles could easily be obtained by chemical synthesis from a mixture of aldehyde-cyanide-ammonia (Strecker reaction).

$$R{-}CHO + CN^- + NH_3 \rightleftharpoons RCH(NH_2)CN \xrightarrow{+H_2O} RCH(NH_2)CONH_2$$

Also, some α-aminonitriles are synthesized and hydrolyzed into L-α-aminoacids by a basidiomycete [151,152] and by *Rhizoctonia solani* [153]; racemic α-aminonitrile precursors of alanine and valine are converted into these amino acids in their racemic form by *Corynebacterium* cells [154].

In the case of the *Brevibacterium* sp. R312 strain, it was necessary to eliminate the wide spectrum amidase activity by a mutation. Such a mutant strain was obtained. The *Brevibacterium* sp. A4 [155–157] retains the wide spectrum nitrile hydratase activity [119,158] and the L-α-aminoamidase activity. This strain is able to transform DL-α-aminonitrile or DL-α-aminoamide into a mixture of D-α-amino amide and L-α-aminoacid. The optically active orientation of the molecule occurs at the amidase level.

The preparation of L-methionine is discussed here as an example. The *Brevibacterium* sp. A4 mutant strain was grown on the YMPG medium. The cells were harvested during the log phase, centrifuged and washed with saline water. The cells were then dispersed in the reaction mixture containing α-amino-γ-methylthiobutyronitrile chlorhydrate 6%. The pH of the mixture was adjusted to between 6.5 and 8.5 by the addition of potassium hydroxide. The bacterial cells (20–40 g l^{-1} dry matter quantitatively transformed the mixture into L-methionine and D-methionine amide (1:1) (1.1).

The products were separated by well-known techniques: crystallization of methionine at pH 7 in the supernatant following elimination of the cells by centrifugation or chromatography on ion exchange resins. The D-amide could be transformed into the D-acid using the amidase of the *Brevibacterium* sp. R312 wild type strain.

The hydrolysis of the nitrile into the L-α-aminoacid could also be performed with acellular preparations. This method is not specific for methionine, and has been successfully applied to the production of L-α-alanine, L-α-aminobutyric acid, L-phenylalanine, L-valine, and L-leucine. It is already possible to apply this technique to the production of other aminoacids such as L-glutamic acid, L-glutamine, and L- and D-phenylglycine from α-aminonitriles already described in the literature.

Finally, the preparation of L-α-aminoacid could be performed from chemically produced DL-α-aminoamide. A general scheme for the production of optically active α-aminoacids has been proposed [159] (Fig. 7).

A number of remarks need to be made. First, through the study of regulation mechanisms and the selection of mutant strains, the production of L-α-aminoamidase could be improved. The selection of nitrile hydratase hyperproducing strains would also improve the first part of the biotransformation. The different properties (pH, optimal temperature in particular) of the two enzymes involved constitute a drawback if only one reactor is used instead of two successive reactors. In any case, immobilization of whole cells or of the enzyme systems would improve the yield of the process. Several problems would remain unsolved:

— only 50% transformation yield,
— difficult separation of the L-α-aminoacid from the D-α-amino amide,
— racemization of the D-α-aminoamide.

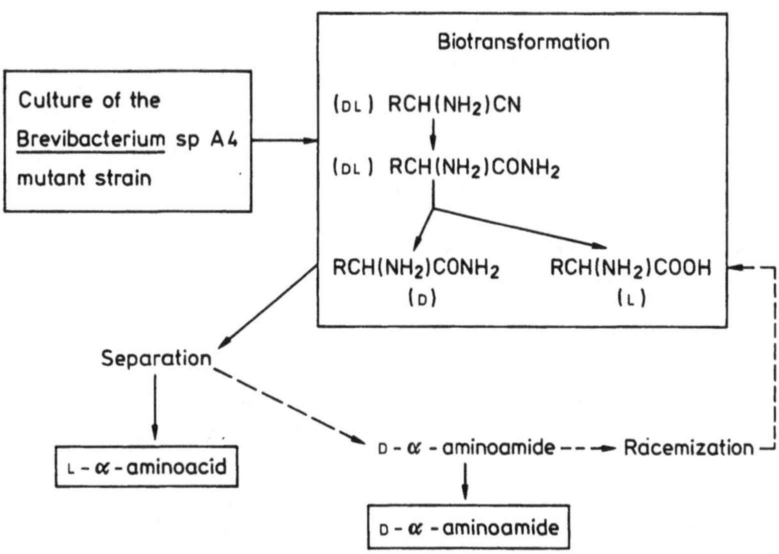

Fig. 7. Scheme for the preparation of optically active α-amino acids

In fact, the last point is the most important in the whole process. If a method could be developed whereby the non-transformed D-amide were continuously racemized, the process would become very effective in transforming the substrate 100% into L-α-amino acid. By the same token, the problem of separating D-α-aminoamide and L-α-aminoacid would no longer exist. The resulting L-α-amino acid as the ammonium salt would then be the sole product in an aqueous solution.

11.2 Production of Organic Acids

11.2.1 Production of Lactic Acid

We chose the production of lactic acid [160] as an example because ammonium lactate possesses great industrial interest [161].

The wide spectrum amidase contributes to the transformation of lactonitrile into DL-ammonium lactate via lactamide as the intermediary step. There is no optical orientation of the molecule.

The wild type *Brevibacterium* sp. R312 was grown on a medium containing glucose as the carbon source. The cells were harvested during the log phase, centrifuged, washed with saline water and dispersed in the reaction mixture which contained 10% lactonitrile. The substrate lactonitrile is commercially available; it could also be produced in situ by reacting an aqueous solution of acetaldehyde with an aqueous solution of cyanhydric acid. The pH of this mixture must be adjusted to ca. 5 with concentrated ammonia for the reaction to proceed. In both cases, the lactonitrile solution was adjusted to pH 7 with ammonia before the addition of the cells. The bacterial cells (20–40 g dry matter per liter) completed the hydrolysis of the nitrile within 2–3 h at 25 °C with agitation. The cells are then removed by centrifugation.

The resulting ammonium lactate in the supernatant could then be recovered by drying. The lactate could also be recovered by other known means, for example, following acidification a continuous extraction could be performed with ethyl ether or any other appropriate organic solvent.

It is noteworthy that under the above conditions the useful life of the cells was relatively long. Experimental results showed that they could be re-employed several times without loss of activity. It is also probable that whole-cell immobilization techniques would improve the cells' useful half-life.

The same process could be used for the production of different acids: DL-α-amino acids, DL-α-hydroxy acids, aliphatic acids, etc. In some cases, the transformation was not completed in 3 hours. This was the case for β-amino propionitrile and nitriles (nicotinonitrile, adiponitrile, . . .) that gave rise to amides not very soluble in water. As for β-aminonitrile, the first step of the process was very rapid, however the hydrolysis of β-aminoamide was slow. The process could be improved by two developments. First, the cells should contain a maximum amount of nitrile hydratase so that the first step of the reaction occurs as rapidly as possible. This could be achieved by the selection of mutant strains that are hyperproducers of nitrile hydratase. An optimization of culture media and conditions would also improve this step. Secondly, the hydrolysis step involving the inducible wide spectrum amidase could be improved by the selection of mutant strains constitutive for this enzyme. Meanwhile, the production of amidase could be improved by adding an inducer such as *N*-methylacetamide to the culture medium at 10 mM. The biosynthesis of the wide spectrum amidase would be induced and would considerably increase so that the second step, amide → acid, would no longer be a limiting factor.

11.2.2 Production of ^{13}C-Marked Keto-acids

A number of laboratories, and especially that of Professor H. L. Schmidt (Lehrstuhl für allgemeine Chemie und Biochemie Technische Universität — München), are attempting to synthesize L-α-aminoacids marked with stable non-radioactive isotopes at different sites. The synthesis of these molecules is interesting since they are used in the study of some metabolic illnesses in man and in the study of alkaloid synthesis by plants.

One of the sites to be marked is the carboxylic group. The non-radioactive isotope used is ^{13}C, which is added as marked potassium cyanide. The reactions involved in the synthesis of marked L-α-aminoacids can be summarized as follows:

$$R-\underset{\underset{O}{\|}}{C}-Cl + K^{13}CN \xrightarrow{(1)} R-\underset{\underset{O}{\|}}{C}-{}^{13}CN \xrightarrow{(2)} R-\underset{\underset{O}{\|}}{C}-{}^{13}COOH \xrightarrow{(3)} R-\underset{\underset{\underset{(L)}{NH_2}}{|}}{CH}-{}^{13}COOH$$

where (1) indicates a chemical reaction, (2) hydrolysis of the nitrile into the acid and (3) is an enzymatic transamination.

The step (2) could perhaps be performed under the joint action of the nitrile hydratase and amidase of *Brevibacterium* sp. R312. These two enzymes have very wide activity spectra, and the biological hydrolysis process would occur under mild conditions as far as pH and temperature are concerned. This would avoid secondary reactions and decomposition of the main substrate, which is a relatively fragile molecule.

In this study the *Brevibacterium* sp. R312 strain was used. Its wide spectrum amidase was induced by *N*-methylacetamide. As marked potassium cyanide is very expensive, the bioconversion experiments were performed with five non-marked α-keto nitriles:

- ketoisocapronitrile $(CH_3)_2CH_2-CO-CN$
- keto-β-chloropropionitrile $ClCH_2-CO-CN$
- ketoisovaleronitrile $(CH_3)_2CH-CO-CN$
- ketopropionitrile $CH_3-CO-CN$
- keto-β-phenylpropionitrile $\Phi-CH_2-CO-CN$

The bioconversion of these nitriles was monitored by GLC. The nitriles were diluted in distilled water, the pH was adjusted to 7 and the concentration was adjusted to 100 mM. The wild type *Brevibacterium* sp. R312 cells were immobilized in calcium alginate gel beads which were added to the reaction mixture. The proportions chosen were 1 g cells (wet weight) for 20 ml nitrile solution (100 mM). The reaction was performed at 20 °C with agitation.

The bioconversion of α-ketopropionitrile into pyruvic acid was especially closely monitored. The first step of the bioconversion (nitrile → amide) was previously tested with the *Brevibacterium* sp. A4 mutant strain defective for the wide spectrum amidase. No α-keto acid could be detected. The second step (amide → acid) was performed with the wild type strain which had its amidase induced. The α-ketopropionitrile was shown to be 100% hydrolyzed into pyruvic acid using commercially available pyruvic acid as standard for the assay. The hydrolysis of the other nitriles was expected to be just as efficient, however, this could not be verified since the corresponding acids are not commercially available to serve as standards for an assay. Note that the reaction rate was much slower for α-keto-β-phenylpropionitrile. Considering the structure of the molecule, this could be due to problems related to diffusion through the cell membrane. This difficulty could eventually be solved by immobilizing the two enzymes by adsorption onto an appropriate carrier and eluting the nitrile solution through a reactor packed with this activated carrier.

Once the bioconversion was completed, the alginate beads were removed from the reaction mixture. The solution was adjusted to pH 9 in order to prevent the acids from becoming too volatile. Following freezing, the solution was lyophilized. Another final treatment option would be to extract the resulting α-keto acids with butanone. This treatment was not retained as losses of the product occurred.

The bioconversion was shown not to reach 100% yield when the nitriles were either not very stable or impure. The decomposition of such compounds often leads to the liberation of cyanide. Although the wide spectrum amidase is not sensitive to cyanide [140], the nitrile hydratase is strongly inhibited by this toxic compound which has an inhibition constant of 4 mM for this enzyme [162]. The process could be improved by the selection of cyanide-resistant strains or by chemically stabilizing the ketonitrile in order to avoid the presence of cyanide (very pure keto-nitriles).

11.2.3 Bioconversion of α-Aminonitriles into Their Corresponding Amino Acids [163]

Considering the considerable adaptation abilities of bacteria and other microorganisms to produce resistance against antibiotics and antifungi drugs, and considering the ever-increasing needs of the health services in order to face this problem, there must be a rational approach to the search for new active compounds in antimicrobial therapeutic applications. Rhizobitoxin is a compound being studied in such an approach [164]. This phytotoxin is produced by a nitrifying bacterium, *Rhizobium japonicum* [165], which fixes nitrogen from the atmosphere into nodules of some soya cultivars and simultaneously induces chloroses of new leaves by the toxin secreted.

Previous studies had shown that this toxin acts on β-cystathionase, an enzyme involved in the biosynthesis of methionine which constitutes a true metabolic crossroad. This sulfur-containing amino acid not only leads to the synthesis of proteins, of *S*-adenosylmethionine and of polyamines, it is also involved in the production of ethylene, a very important plant hormone. As this enzyme modulates or suppresses the action of ethylene, its action could result in the retardation of the flowering of buds or of ripening of green fruits, the slowing of the senescence of flowers and ripe fruits, and an increase of the longevity of cut flowers. It could affect the regulation and control of the germination process. Potential applications of the knowledge and understanding of this compound are considerable and far reaching.

There are only two known analogues of the rhizobitoxin:

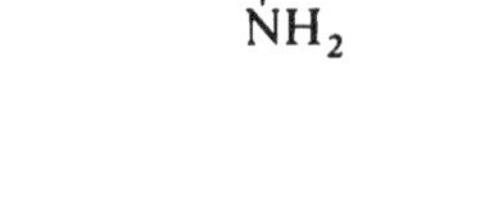

— methoxyvinylglycine R = CH_3
— aminoethoxyvinylglycine R = $CH_2—CH_2—NH_2$
Similarly to rhizobitoxin, these two compounds inhibit the biosynthesis of proteins and ethylene.

Workers from the C.E.R.C.O.A. (Centre d'Etudes et de Recherches de Chimie Organique Appliquée) and of the Laboratoire de Bioorganique et Biotechnologie of the E.N.S.C.P. (Ecole Nationale Supérieure de Chimie de Paris) under the direction of Professeur F. Le Goffic have been trying to synthesize other structural analogues of rhizobitoxin [166,167]. The aim of these researchers was to determine which are the functional groups responsible for the above properties. By modifying the structure of the "leader" molecule, it is possible to determine whether a particular structural change will affect the biological properties of the resulting compound or not.

The synthesis of such complex α-amino acids is achieved through the elaboration of their corresponding nitriles. However, chemical methods for the hydrolysis of these precursors are most often inapplicable for R residues with fragile polyfunctional groups. The bioconversion of nitriles into acids by the *Brevibacterium* sp. R312 appeared to be a mild enough method to be used in this organic synthesis procedure.

The bioconversion of these very special nitriles was performed in two steps. The first part involved the wide substrate spectrum nitrile hydratase. The second step could be achieved either with the wide spectrum amidase, resulting in a racemic mixture of amino acids, or with the L-α-amino amidase producing a mixture of D-amide and L-amino acid. The first results obtained showed that this type of bioconversion by a bacterial strain with a nitrile hydratase and an amidase could have many applications in organic chemical syntheses.

The bioconversion was performed with the wild type *Brevibacterium* sp. R312, the amidase of which was induced by *N*-methylacetamide. The bioconversion tests were run with 4 α-aminonitriles with the following formulas:

$$R-O-CH_2-CH_2-\underset{\displaystyle NH_2}{\underset{|}{CH}}-CN$$

The R radical represents in
nitrile A: $CH_3—$,
nitrile B: $CH_3—CH_2—$,
nitrile C: $(CH_3)_2—CH_2—$,
nitrile D:

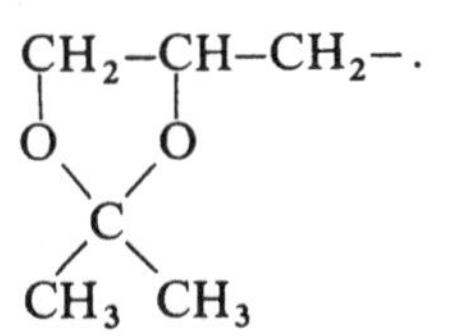

First 1.5 g of each nitrile was dissolved in 50 ml phosphate buffer (200 mM; pH 7). Then 0.5 g of bacterial cells with nitrile hydratase and amidase activities determined to be respectively 200 μmoles propionitrile and 20 μmoles propionamide per minute per mg protein was dispersed in 50 ml of the same phosphate buffer. The cell suspension

was mixed with the nitrile solution. The reaction was allowed to proceed for 4 h at 28 °C with agitation. Aliquots were taken during the bioconversion process and deposited on TLC plates. Following migration in the propanol:ammonia 20% (70:30) solvent system, the different α-amino compounds were revealed by spraying with ninhydrin. Reaction time was found to be insufficient for all four nitriles tested. The cells were removed by centrifugation at 10000 rpm for 10 min and replaced by fresh cells. The bioconversion was resumed for another 4 h. It was already apparent that the bioconversion rates were slower when the molecules were larger and spatially crowded. This seemed to be tied to the problem of penetration by the nitriles into the cells. Following a total of 8 h bioconversion, nitriles A, B, and C were completely hydrolyzed into their corresponding acids. Such was not the case for nitrile D.

In order to facilitate the access of nitrile D to the active sites of both enzymes, the reaction was completed with a sonicated acellular preparation. Using 1 volume of enzyme preparation for 5 volumes of substrate, the bioconversion was complete in 2.5 h. The proteins were precipitated out by the addition of a saturated ammonium sulfate solution. The four nitriles thus prepared were lyophilized for storage. They could be purified by eluting a solution through a Dowex 50×2 — 200, a cationic resin.

A colored spot on the TLC was revealed by ninhydrin when non-immobilized whole cells of *Brevibacterium* sp. R312 were used. This spot probably corresponds to amino compounds liberated by the cells during the bioconversion process. The effect of cell immobilization in calcium alginate on this coloration was verified using nitrile A as the substrate. The results showed that when the cells were immobilized the metabolite liberation problem was drastically reduced or even completely eliminated.

The mild technique described above could be used for polyfunctional molecules which would be unstable in the usual chemical reaction conditions. This efficient and elegant bioconversion process could be very important and useful in the synthesis of organic compounds, especially if it were enantioselective.

11.2.4 Production of Acrylic Acid [160]

Acrylic acid may be produced from either acrylonitrile or acrylamide. Acrylamide itself may be produced by chemical hydratation [168] or biological hydratation [169-177]. Acrylic acid is a commercially very important compound [178] and is currently produced by large chemical concerns. The fixed bed reactors previously described (see Sects. 11.2.2 and 11.2.3) were not able to be used for the present process. Instead, a multiphasic reactor was designed [179,180]. This is a fluidized bed reactor in which three phases were mixed. The three phases are: air, the aqueous solution containing the substrate and/or the product, and the carrier particles chosen for immobilizing the biocatalyst.

In this experiment calcium alginate beads 3 mm in diameter occupying a volume of about 290 ml and containing 15 g (wet weight) of the *Brevibacterium* sp. R312 cells were used. A 100% yield was obtained in the processing of a 1% acrylamide solution at the rate of 175 ml h^{-1}.

These first results were encouraging as under some operating conditions the multiphasic reactor performed the acrylamide-acrylate bioconversion with very little energetic requirement [181]. In any case, the fluidized bed reactor was shown to be far superior to the fixed bed reactor. Besides, new designs are already being considered in order to improve the operation of the reactor:

a) A decrease in the size of the carrier particles. This would increase the hydratation rate for the same number of immobilized cells.

b) An increase in the total charge of the solid carrier.

c) Elimination of the aeration technique, which oxidized the thiol groups of the amidase causing accelerated denaturation of the enzyme. Fluidization of the particles could be achieved instead by pumping through the recycled liquid phase.

12 Conclusion and Perspective

In the present paper we have reviewed the enzyme systems involved in the hydrolysis of amides and the analytical techniques used in their study. Applications of this type of activity were also discussed. The main feature of interest in amidases of *Brevibacterium* sp. R312 was their involvement as an essential step in the biological hydrolysis of nitriles. This bacterial strain has a wide spectrum nitrile hydratase. Consequently, it is possible to hydrolyze fragile nitrile compounds into the corresponding acids with the wide spectrum amidase or, by using the L-α-aminoamidase instead, optical specificity could be obtained for the resulting acids.

The knowledge assembled to date would be very useful for improving the amide-acid step so that this would no longer be the limiting factor in the bioconversion of nitriles into acids. Several points have to be considered:

a) Optimization of the two amidases (wide spectrum amidase, L-α-aminoamidase) by modification of pH, temperature, activating cations, medium composition, etc.

b) Genetic improvement of the strain. This improvement could be achieved by two methods:

• Classical genetics. This would involve the selection of constitutive mutant hyperproducers of the wide spectrum amidase and derepressed mutant hyperproducers of L-α-aminoamidase, using the recent findings on the regulation of the biosynthesis of these enzymes.

• Genetic engineering. Attempts at cloning the genes corresponding to the wide spectrum amidase and to the L-α-aminoamidase are in progress. If successful, the resulting genes will be used for expression amplification tests.

c) New techniques for the immobilization of whole cells and enzymes. Covalent bonding will be particularly studied.

d) Development of a fluidized bed reactor. Such a reactor would be useful for the transformation of large quantities of product with a low space requirement.

New applications for the bioconversion of nitriles and amides will be found as improvements of the present process are made. These applications could be in the fields of organic chemistry, water treatment, or the food industry.

13 References

1. Jallageas, J. C., Arnaud, A., Galzy, P.: Adv. Biochem. Eng. *14*, 1 (1980)
2. Thimann, K. V., Mahadevan, S.: Nature *181*, 1466 (1958)
3. Thimann, K. V., Mahadevan, S.: Arch. Biochem. Biophys. *105*, 133 (1963)

4. Mahadevan, S., Thimann, K. V.: ibid. *107*, 62 (1964)
5. Harper, D. B.: Biochem. Soc. Trans. *4*, 502 (1976)
6. Harper, D. B.: Biochem. J. *165*, 309 (1977)
7. Harper, D. B.: ibid. *167*, 685 (1977)
8. Smith, E. L., Hill, R. L.: Enzymes *4*, 37 (1960)
9. Smith, E. L., Polglase, W. J.: J. Biol. Chem. *180*, 1209 (1949)
10. Yellin, T. O., Wriston, J. C.: Biochemistry *5*, 1605 (1966)
11. Ramadan, M. E. A., El Asmar, E., Greenberg, D. M.: Arch. Biochem. Biophys. *108*, 143 (1964)
12. Meister, A., Levintow, L., Greenfield, R. E., Abendschein, P. A.: J. Biol. Chem. *215*, 441 (1955)
13. Fernald, N. J., Ramaley, R. F.: Arch. Biochem. Biophys. *153*, 95 (1972)
14. Larson, A. D., Kallio, R. E.: J. Bacteriol. *68*, 67 (1954)
15. Lister, A. J.: J. Gen. Microbiol. *14*, 478 (1956)
16. Koivusalo, M., Elorriaga, C., Kaziro, Y., Ochoa, S.: J. Biol. Chem. *238*, 1038 (1963)
17. Joshi, J. G., Handler, P.: ibid. *237*, 929 (1962)
18. Kimura, T.: J. Biochem. *46*, 973 (1959)
19. Kimura, T.: ibid. *46*, 1133 (1959)
20. Kimura, T.: ibid. *46*, 1271 (1959)
21. Kimura, T.: ibid. *46*, 1399 (1959)
22. Reitz, M. S., Rodwell, V. W.: J. Biol. Chem. *245*, 3091 (1970)
23. Vogels, G. D., Trijbels, F., Uffink, A.: Biochim. Biophys. Acta *122*, 482 (1966)
24. Berger, J., Johnson, M. J., Peterson, W. H.: J. Bacteriol. *36*, 521 (1938)
25. Brown, J. L.: J. Biol. Chem. *248*, 409 (1973)
26. Chapuis, R., Zuber, H.: Methods Enzymol. *19*, 552 (1970)
27. De Marco, A. C., Dick, A. J.: Can. J. Biochem. *56*, 66 (1978)
28. Foissy, H.: Milchwissenschaft *33*, 221 (1978)
29. Foissy, H.: Z. Lebensm.-Unters.-Forsch. *166*, 164 (1978)
30. Hayman, S., Gatmaitan, J. S., Patterson, E. K.: Biochemistry *13*, 4486 (1974)
31. Ivanova, N. M., Vaganova, T. I., Strongin, A. Y., Stepanov, V. M.: Biochemistry (USSR) *42*, 652 (1977)
32. Koelsch, R., Hanson, H.: Acta Biol. Med. Ger. *26*, 1117 (1971)
33. Lehmann, K., Uhlig, H.: Hoppe-Seyler's Z. Physiol. Chem. *350*, 99 (1969)
34. Masuda, T. M., Hayashi, R., Hata, T.: Agric. Biol. Chem. *39*, 499 (1975)
35. Matheson, A. T.: Can. J. Biochem. *41*, 8 (1963)
36. Matheson, A. T., Bjerre, S., Hanes, C. S.: Can. J. Biochem. Physiol. *41*, 1741 (1963)
37. Matheson, A. T., Tsai, C. S.: Can. J. Biochem. *43*, 323 (1965)
38. Tsai, C. S., Matheson, A. T.: ibid. *43*, 1643 (1965)
39. Mimamiura, N., Yamamoto, T., Fukumoto, J.: Agric. Biol. Chem. *30*, 186 (1966)
40. Mimamiura, N., Matsumura, Y., Yamamoto, T., Fukumoto, J.: ibid. *33*, 653 (1969)
41. Morihara, K., Tsuzuki, H.: Ann. Rep. Shionogi Res. Lab. *13*, 91 (1963)
42. Nakadai, T., Nasuno, S., Iguchi, N.: Report of the Noda Institute for Scientific Research *18*, 16 (1974)
43. Nakadai, T., Nasuno, S., Iguchi, N.: Agric. Biol. Chem. *41*, 25 (1977)
44. Nakadai, T., Nasuno, S., Iguchi, N.: ibid. *41*, 33 (1977)
45. Nakadai, T., Nasuno, S., Iguchi, N.: ibid. *41*, 1657 (1977)
46. Prescott, J. M., Wilkes, S. H.: Meth. Enzymol. *45* B, 530 (1976)
47. Roncari, G., Zuber, H.: Meth. Enzymol. *19*, 544 (1970)
48. Roncari, G., Stoll, E., Zuber, H.: Meth. Enzymol. *45* B, 522 (1976)
49. Ruffin, P., Vanbrussel, E., Biguet, J., Biserte, G.: Biochimie *61*, 495 (1979)
50. Sugiura, M., Suzuki, M., Ishikawa, M., Sasaki, M.: Chem. Pharm. Bull. *24*, 2286 (1976)
51. Sugiura, M., Suzuki, M., Ishikawa, M., Sasaki, M.: ibid. *24*, 2026 (1976)
52. Sugiura, M., Ishikawa, M., Sasaki, M.: ibid. *26*, 3101 (1978)
53. Uwajima, T., Yoshikawa, U., Osamu, T.: Agric. Biol. Chem. *37*, 2727 (1973)
54. Vogt, V. M.: J. Biol. Chem. *245*, 4760 (1970)
55. Wagner, F. N., Chung, A., Ray, L. E.: Can. J. Microbiol. *18*, 1883 (1972)
56. Clarke, P. H.: Adv. Microbiol. Physiol. *4*, 179 (1970)
57. Gorr, G., Wagner, J.: Biochem. Z. *266*, 96 (1933)
58. Hynes, M. J., Pateman, J. A.: J. Gen. Microbiol. *63*, 317 (1970)

59. Hynes, M. J.: J. Bacteriol. *103*, 482 (1970)
60. Hynes, M. J.: J. Gen. Microbiol. *91*, 99 (1975)
61. Grant, D. J. W.: Antonie van Leeuwenhoeck; J. Microbiol. Serol. *39*, 273 (1973)
62. Grant, D. J. W., Wilson, J. V.: Microbios *8*, 15 (1973)
63. Halpern, Y. S., Grossowicz, N.: Biochem. J. *65*, 716 (1957)
64. Draper, P.: J. Gen. Microbiol. *46*, 111 (1967)
65. Viallier, J., Viallier, G.: Rev. Inst. Pasteur Lyon *4*, 167 (1971)
66. Georges, J. C., Dailloux, M.: Ann. Biol. Clin. (Paris) *31*, 217 (1973)
67. Kelly, M., Kornberg, H. L.: J. Biochem. *93*, 557 (1964)
68. Jakobi, W. B., Fredericks, J.: J. Biol. Chem. *239*, 1978 (1964)
69. Clarke, P. H.: Biochem. Soc. Trans. (549th meeting, Cambridge) *2*, 831 (1974)
70. Betz, J. L., Brown, P. R., Smyth, P. R., Clarke, P. H.: Nature *247*, 261 (1974)
71. Firmin, J. L., Gray, D. O.: Biochem. J. *158*, 223 (1976)
72. Arnaud, A., Galzy, P., Jallageas, J. C.: Folia Microbiol. (Prague) *21*, 178 (1976)
73. Thalenfeld, B., Grossowicz, N.: J. Gen. Microbiol. *94*, 131 (1976)
74. Di Geronimo, M. J., Antoine, A. D.: Appl. Environ. Microbiol. *31*, 900 (1976)
75. Hagikara, B.: Enzymes *4*, 193 (1960)
76. Asano, Y., Tachibana, M., Tani, Y., Yamada, H.: Agric. Biol. Chem. *46*, 1175 (1982)
77. Miller, J. M., Gray, D. O.: J. Gen. Microbiol. *128*, 1803 (1981)
78. Friedrich, C. G., Mitrenga, G.: ibid. *125*, 367 (1981)
79. Hynes, M. J.: J. Bacteriol. *131*, 770 (1977)
80. Hynes, M. J.: ibid. *142*, 400 (1980)
81. Hynes, M. J.: Mol. Gen. Genet. *161*, 59 (1978)
82. Hynes, M. J.: ibid. *166*, 31 (1978)
83. Hynes, M. J.: Genetics *91*, 381 (1979)
84. Hynes, M. J., Pateman, J. A.: Mol. Gen. Genet. *108*, 107 (1970)
85. Dunsmuir, P., Hynes, M. J.: ibid. *123*, 333 (1973)
86. Hynes, M. J.: Nature *253*, 210 (1975)
87. Apirion, D.: Genet. Res. *6*, 317 (1965)
88. Armitt, S., MacCullough, W., Roberts, C. F.: J. Gen. Microbiol. *92*, 263 (1976)
89. Arst, H. N.: Nature *262*, 231 (1976)
90. Arst, H. N., Cove, D. J.: Mol. Gen. Genet. *126*, 111 (1973)
91. Hynes, M. J.: ibid. *125*, 99 (1973)
92. Hynes, M. J.: Aust. J. Biol. Sci. *28*, 301 (1975)
93. Clarke, P. H.: J. Gen. Microbiol. *71*, 241 (1972)
94. Betz, J. L., Clarke, P. H.: ibid. *75*, 167 (1973)
95. Kelly, M., Clarke, P. H.: ibid. *27*, 305 (1962)
96. Brammar, W. J., Clarke, P. H.: ibid. *37*, 307 (1964)
97. Brammar, W. J., MacFarlane, N. D., Clarke, P. H.: ibid. *44*, 303 (1966)
98. Boddy, A., Clarke, P. H., Houldsworth, M. A., Lilly, M. D.: ibid. *48*, 137 (1967)
99. Clarke, P. H., Houldsworth, M. A., Lilly, M. D.: ibid. *51*, 225 (1968)
100. Brown, P. R., Clarke, P. H.: ibid. *70*, 287 (1972)
101. Brammar, W. J., Clarke, P. H., Skinner, A. J.: ibid. *47*, 87 (1967)
102. Brown, J. E., Brown, P. R., Clarke, P. H.: ibid. *57*, 273 (1969)
103. Brown, J. E., Clarke, P. H.: ibid. *64*, 329 (1970)
104. Clarke, P. H., Tata, R.: ibid. *75*, 231 (1973)
105. Betz, J. L., Clarke, P. H.: ibid. *73*, 161 (1972)
106. Betz, J. L., Brown, P. R., Smyth, P. R., Clarke, P. H.: Nature *247*, 261 (1974)
107. Gregoriou, M., Brown, P. R.: Arch. Microbiol. *125*, 277 (1980)
108. Brown, P. R., Smyth, M. J., Clarke, P. H., Rosemeyer, M. A.: Eur. J. Biochem. *34*, 177 (1973)
109. Paterson, A., Clarke, P. H.: J. Gen. Microbiol. *114*, 75 (1979)
110. Jallageas, J. C., Arnaud, A., Galzy, P.: J. Chromatogr. *166*, 181 (1978)
111. Jallageas, J. C., Arnaud, A., Galzy, P.: Anal. Biochem. *95*, 436 (1979)
112. Bui, K., Fradet, H., Maestracci, M., Thiéry, A., Arnaud, A., Galzy, P.: Third European Congress of Biotechnology (Verlag Chemie — Dechema) *1*, 415 (1984)
113. Muftic, M. K.: Nature *20*, 623 (1964)
114. Mimura, A., Kawano, T., Yamada, K.: J. Ferment. Technol. *47*, 631 (1969)

115. Conway, E. J., Byrne, A.: Biochem. J. *27*, 419 (1933)
116. Conway, E. J., O'Brien, M., Boyle, P. J.: Nature *148*, 662 (1941)
117. Arnaud, A., Galzy, P., Jallageas, J. C.: C.R. Acad. Sci. *283*, 571 (1976)
118. Arnaud, A., Galzy, P., Jallageas, J. C.: Rev. Ferment. Ind. Aliment. *31*, 39 (1976)
119. Bui, K., Fradet, H., Arnaud, A., Galzy, P.: J. Gen. Microbiol. *130*, 89 (1984)
120. Thiéry, A., Maestracci, M., Arnaud, A., Galzy, P.: Zbl. Mikrobiol. *141*, 575 (1986)
121. Gilbert, T. R., Clay, A. M.: Anal. Chem. *45*, 1575 (1973)
122. Thomas, R. F., Booth, R. L.: Environ. Sci. Technol. *7*, 523 (1973)
123. Eagan, M. L., Dubois, L.: Anal. Chim. Acta *70*, 157 (1974)
124. Mertens, J., Van der Winkel, P., Massart, D. L.: Bull. Soc. Chim. Belg. *83*, 19 (1974)
125. Attili, A. F., Autizi, D., Capocaccia, L.: Biochem. Med. *14*, 109 (1975)
126. Shibata, N.: Anal. Chim. Acta *83*, 371 (1976)
127. Kieny-L'Homme, M. P., Arnaud, A., Galzy, P.: J. Gen. Appl. Microbiol. *27*, 307 (1981)
128. Schneider, F., Lefebvre, G., Gay, R., Raval, G.: Biochimie *60*, 45 (1978)
129. Joseph, R. L., Sanders, W. J.: Biochem. J. *100*, 827 (1966)
130. Wachsmuth, E. D., Fritze, I., Pfleiderer, G.: Biochemistry *5*, 169 (1966)
131. Hanson, H., Hutter, H. J., Mannsfeld, H. G., Kretschmer, K., Sohr, C.: Hoppe-Seyler's Z. Physiol. Chem. *348*, 680 (1967)
132. Brecher, A. S., Suszkiw, J. B.: Biochem. J. *112*, 335 (1969)
133. Charet, P., Aissi, E., Maurois, P. Bouquelet, S., Biguet, J.: Comp. Biochem. Physiol. *65B*, 519 (1980)
134. Miller, J. M., Knowles, C. J.: FEMS Microbiol. Lett. *21*, 147 (1984)
135. Thiéry, A., Maestracci, M., Arnaud, A., Galzy, P., Nicolas, M.: J. Basic Microbiol. *26*, 299 (1986)
136. Brown, P. R., Smyth, M. J., Clarke, P. H., Rosemeyer, M. A.: Eur. J. Biochem. *34*, 177 (1973)
137. Laemmli, V. K.: Nature *227*, 680 (1970)
138. Thiéry, A., Maestracci, M., Arnaud, A., Galzy, P.: J. Gen. Microbiol. *132*, 2205 (1986)
139. Maestracci, M., Thiéry, A., Arnaud, A., Galzy, P.: Agric. Biol. Chem. *50*, 2237 (1986)
140. Maestracci, M., Thiéry, A., Bui, K., Arnaud, A., Galzy, P.: Arch. Microbiol. *138*, 315 (1984)
141. Thiéry, A., Maestracci, M., Arnaud, A., Galzy, P.: Belgian J. Food Chem. Biotechnol. *40*, 115 (1985)
142. Magasanik, B.: Cold Spring Harbor Symp. Quant. Biol. *26*, 249 (1961)
143. Zimmermann, F. K., Schell, I.: Mol. Gen. Genet. *154*, 75 (1977)
144. Cove, D. J.: Biochim. Biophys. Acta *113*, 51 (1966)
145. Kinsky, S. C., MacElroy, W. D.: Arch. Biochem. Biophys. *73*, 466 (1958)
146. Thiéry, A.: Thèse de Docteur-Ingénieur — ENSA — Montpellier (1985)
147. Tourneix, D., Thiéry, A., Maestracci, M., Arnaud, A., Galzy, P.: Antonie van Leeuwenhoek; J. Microbiol. Serol. *52*, 173 (1986)
148. Willhite, C. C., Ferm, V. H., Smith, R. P.: Teratology *23*, 317 (1981)
149. Novo Industri. Brevet Lux. 77, 00320 (1977)
150. Takeda, H., Matsumoto, I., Naito, M.: Jap. Patent 78, 66493 (1978)
151. Strobel, G. A.: J. Biol. Chem. *241*, 2618 (1966)
152. Strobel, G. A.: ibid. *242*, 3265 (1967)
153. Mundy, B. P., Lin, I. H. S., Strobel, G. A.: Can. J. Biochem. *51*, 1440 (1973)
154. Fukuda, Y., Fukui, M., Harada, T., Izumi, Y.: J. Ferment. Technol. *51*, 393 (1973)
155. Jallageas, J. C., Arnaud, A., Galzy, P.: C.R. Acad. Sci. *288*, 655 (1979)
156. Jallageas, J. C., Arnaud, A., Galzy, P.: Adv. Biotechnol. *3*, 227 (1981)
157. Arnaud, A., Galzy, P., Jallageas, J. C.: Bull. Soc. Chim. *II*, 87 (1980)
158. Arnaud, A., Jallageas, J. C., Galzy, P.: Agric. Biol. Chem. *41*, 2183 (1977)
159. Galzy, P., Arnaud, A., Jallageas, J. C.: Brevet Fr. 79,01803 (1979)
160. Arnaud, A., Galzy, P., Commeyras, A., Jallageas, J. C.: Brevet ANVAR-CNRS Fr. 73,33613 (1973)
161. Marriott, T. A.: J. Soc. Dairy Technol. *38*, 109 (1985)
162. Bui, K., Maestracci, M., Thiéry, A., Arnaud, A., Galzy, P.: J. Appl. Bacteriol. *57*, 183 (1984)
163. Marrais, D., Vo Quang, L., Vo Quang, Y., Le Goffic, F., Thiéry, A., Maestracci, M., Arnaud, A., Galzy, P.: IVth European Symposium on Organic Chemistry. Aix-en-Provence (France) September 2–6 (1985)

164. Owens, L. D., Thompson, J. F., Pitcher, R. G., Williams, T.: J. Chem. Soc., Chem. Commun. 714 (1972)
165. Owens, L. D., Wright, D. A.: Plant Physiol. *40*, 927 (1965)
166. Vo Quang, Y., Marais, D., Vo Quang, L., Le Goffic, F.: Tetrahedron Lett. *24*, 5209 (1983)
167. Vo Quang, Y., Carniato, D., Vo Quang, L., Le Goffic, F.: J. Chem. Soc., Chem. Commun., 1505 (1983)
168. Matsuda, F., Hashimoto, N.: Inf. Chim. *160*, 223 (1976)
169. Bui, K., Arnaud, A., Galzy, P.: Enzyme Microbiol. Technol. *4*, 195 (1982)
170. Galzy, P., Arnaud, A., Commeyras, A., Jallageas, J. C.: Brevet ANVAR-CNRS 74,41828 (1974)
171. Fradet, H., Arnaud, A., Rios, G., Galzy, P.: Biotechnol. Bioeng. *27*, 1581 (1985)
172. Asano, Y., Yasuda, T., Tani, Y., Yamada, H.: Agric. Biol. Chem. *46*, 1183 (1982)
173. Watanabe, I., Satoh, Y., Takano, T.: Ger. Offen. 2,912,292 (1979)
174. Nitto Chemical Industry Co, Ltd: Brevet Fr. 79,07935 (1979)
175. Nitto Chemical Industry Co, Ltd: ibid. 80,13635 (1980)
176. Nitto Chemical Industry Co, Ltd: ibid. 80,03188 (1980)
177. Nitto Chemical Industry Co, Ltd: ibid. 81,15932 (1981)
178. Watanabe, I.: Jap. Patent 79,46887 (1979)
179. Rios, G. M., Baxerres, J. L., Gibert, H.: Brevet Fr. 79,17430 (1979)
180. Jacquet, M., Vincent, J. C., Rios, G. M., Gibert, H.: Thé, Cacao, Café *1*, 45 (1981)
181. Papaconstantinou, S., Elmaleh, S., Rios, G. M.: International Symposium on "Heat and Mass transfer in fixed and fluidized beds" Dubrovnik (1984)

ISBN 3-05-500568-6